— 150 — NATIONAL PARKS

YOU NEED TO VISIT BEFORE — YOU DIE —

By Bailey Berg

Lannoo

Welcome to a journey through some of our planet's most awe-inspiring landscapes.

From the towering peaks of the Himalayas to the vast expanses of the African savannah and from the ancient forests of North America to the tropical islands of the South Pacific, these parks represent the epitome of natural beauty and wilderness preservation. Each one is a testament to the remarkable diversity and resilience of life on Earth.

The concept of national parks traces its roots back to the 19th century, a time when industrialization and urbanization began to encroach upon once-pristine landscapes. Recognizing the need to protect these natural wonders for future generations, visionary individuals and governments around the world began setting aside areas of land as protected reserves.

In 1872, Yellowstone National Park was established in the United States, marking the birth of the modern national park system and igniting a global movement to conserve and preserve natural landscapes. This groundbreaking initiative inspired countries around the world to follow suit – today there are more than 6,000 worldwide, according to the International Union for Conservation of Nature.

In this guide, we will delve into the wonders of 150 of these extraordinary landscapes, uncovering their rich history, unique ecosystems, and the myriad adventures they offer intrepid travelers. Whether you're a seasoned outdoors enthusiast seeking your next great expedition or a nature lover yearning to connect with the wild, these national parks promise an unforgettable experience.

Bailey Berg

OVERVIEW

AFRICA

ALGERIA
01 **GOURAYA NATIONAL PARK** P.10
BOTSWANA
02 **CHOBE NATIONAL PARK** P.11
DEMOCRATIC REPUBLIC OF THE CONGO
03 **VIRUNGA NATIONAL PARK** P.12
EGYPT
04 **WADI EL GEMAL NATIONAL PARK** P.14
05 **WHITE DESERT NATIONAL PARK** P.15
ETHIOPIA
06 **SIMIEN MOUNTAINS NATIONAL PARK** P.16
KENYA
07 **HELL'S GATE NATIONAL PARK** P.17
08 **LAKE NAKURU NATIONAL PARK** P.18
MADAGASCAR
09 **ISALO NATIONAL PARK** P.19
NAMIBIA
10 **SKELETON COAST NATIONAL PARK** P.20
11 **ETOSHA NATIONAL PARK** P.22
12 **NAMIB-NAUKLUFT NATIONAL PARK** P.23
REPUBLIC OF THE CONGO
13 **ODZALA-KOKOUA NATIONAL PARK** P.24
RÉUNION
14 **RÉUNION NATIONAL PARK** P.25
RWANDA
15 **VOLCANOES NATIONAL PARK** P.28
SENEGAL
16 **NATIONAL PARK LANGUE DE BARBARIE** P.30

SOUTH AFRICA
17 **KRUGER NATIONAL PARK** P.31
18 **TABLE MOUNTAIN NATIONAL PARK** P.34
TANZANIA
19 **SERENGETI NATIONAL PARK** P.35
20 **KILIMANJARO NATIONAL PARK** P.39
21 **NGORONGORO NATIONAL PARK** P.40
UGANDA
22 **BWINDI IMPENETRABLE NATIONAL PARK** P.41
ZAMBIA
23 **SOUTH LUANGWA NATIONAL PARK** P.44
ZIMBABWE
24 **HWANGE NATIONAL PARK** P.46
25 **VICTORIA FALLS NATIONAL PARK** P.47

THE AMERICAS

ARGENTINA/BRAZIL
26 **IGUAZÚ NATIONAL PARK** P.51
ARGENTINA
27 **TIERRA DEL FUEGO NATIONAL PARK** P.52
28 **LOS GLACIARES NATIONAL PARK** P.53
ARUBA
29 **ARIKOK NATIONAL PARK** P.57
BOLIVIA
30 **AMBORÓ NATIONAL PARK** P.58
BRAZIL
31 **CHAPADA DIAMANTINA NATIONAL PARK** P.60
32 **APARADOS DA SERRA NATIONAL PARK** P.64
33 **LENÇÓIS MARANHENSES NATIONAL PARK** P.66

OVERVIEW

34	**FERNANDO DE NORONHA MARINE NATIONAL PARK**	P.67
	CANADA	
35	**JASPER NATIONAL PARK**	P.69
36	**FORILLON NATIONAL PARK**	P.70
37	**BANFF NATIONAL PARK**	P.73
	CHILE	
38	**TORRES DEL PAINE NATIONAL PARK**	P.74
39	**RAPA NUI NATIONAL PARK**	P.77
	COSTA RICA	
40	**TORTUGUERO NATIONAL PARK**	P.78
41	**CORCOVADO NATIONAL PARK**	P.80
	ECUADOR	
42	**GALÁPAGOS NATIONAL PARK**	P.81
43	**COTOPAXI NATIONAL PARK**	P.84
	GUATEMALA	
44	**TIKAL NATIONAL PARK**	P.85
	GUYANA	
45	**KAIETEUR NATIONAL PARK**	P.86
	MEXICO	
46	**TULUM NATIONAL PARK**	P.87
	PANAMA	
47	**VOLCAN BARU NATIONAL PARK**	P.89
	PERU	
48	**MANU NATIONAL PARK**	P.90
	UNITED STATES	
49	**MOUNT RAINIER NATIONAL PARK**	P.92
50	**NEW RIVER GORGE NATIONAL PARK**	P.94
51	**VOYAGEURS NATIONAL PARK**	P.95
52	**WHITE SANDS NATIONAL PARK**	P.98
53	**WRANGELL-ST. ELIAS NATIONAL PARK**	P.99
54	**YELLOWSTONE NATIONAL PARK**	P.102
55	**YOSEMITE NATIONAL PARK**	P.104
56	**ZION NATIONAL PARK**	P.105
57	**PETRIFIED FOREST NATIONAL PARK**	P.106
58	**JOSHUA TREE NATIONAL PARK**	P.108
59	**GLACIER NATIONAL PARK**	P.109
60	**GRAND CANYON NATIONAL PARK**	P.110
61	**KATMAI NATIONAL PARK**	P.112
62	**KENAI FJORDS NATIONAL PARK**	P.113
63	**GRAND TETON NATIONAL PARK**	P.115
64	**GREAT SMOKY MOUNTAINS NATIONAL PARK**	P.116
65	**HALEAKALĀ NATIONAL PARK**	P.117
66	**HAWAI'I VOLCANOES NATIONAL PARK**	P.118
67	**DENALI NATIONAL PARK**	P.120
68	**BIG BEND NATIONAL PARK**	P.121
69	**REDWOOD NATIONAL PARK**	P.122
70	**ROCKY MOUNTAIN NATIONAL PARK**	P.125
71	**SEQUOIA AND KINGS CANYON NATIONAL PARK**	P.126
72	**SHENANDOAH NATIONAL PARK**	P.128
73	**ARCHES NATIONAL PARK**	P.129
74	**ACADIA NATIONAL PARK**	P.132
75	**CHANNEL ISLANDS NATIONAL PARK**	P.134
	VENEZUELA	
76	**CANAIMA NATIONAL PARK**	P.135

ASIA

	BHUTAN	
77	**JIGME DORJI NATIONAL PARK**	P.138
	BORNEO	
78	**KINABALU NATIONAL PARK**	P.139
79	**NIAH NATIONAL PARK**	P.141
80	**LAMBIR HILLS NATIONAL PARK**	P.142

OVERVIEW

81	**GUNUNG MULU NATIONAL PARK**	P.143
	CAMBODIA	
82	**PREAH MONIVONG BOKOR NATIONAL PARK**	P.144
	CHINA	
83	**ZHANGYE DANXIA NATIONAL GEOPARK**	P.145
84	**BADALING NATIONAL FOREST PARK**	P.146
85	**GUILIN AND LIJIANG RIVER NATIONAL PARK**	P.147
	ISRAEL	
86	**MASADA NATIONAL PARK**	P.150
	INDIA	
87	**SUNDARBANS NATIONAL PARK**	P.152
88	**RANTHAMBORE NATIONAL PARK**	P.153
	JAPAN	
89	**FUJI-HAKONE-IZU NATIONAL PARK**	P.154
	JORDAN	
90	**WADI RUM PROTECTED AREA**	P.155
	INDONESIA	
91	**KOMODO NATIONAL PARK**	P.158
	KAZAKHSTAN	
92	**ALTYN-EMEL NATIONAL PARK**	P.160
	MONGOLIA	
93	**GORKHI-TERELJ NATIONAL PARK**	P.161
	NEPAL	
94	**SAGARMATHA NATIONAL PARK**	P.162
95	**CHITWAN NATIONAL PARK**	P.163
	PHILIPPINES	
96	**HUNDRED ISLANDS NATIONAL PARK**	P.164
	RUSSIA	
97	**STOLBY NATURE RESERVE**	P.165
	SOUTH KOREA	
98	**SEORAKSAN NATIONAL PARK**	P.166
99	**NAEJANGSAN NATIONAL PARK**	P.167
100	**JIRISAN NATIONAL PARK**	P.168
	SRI LANKA	
101	**YALA NATIONAL PARK**	P.169
	TAIWAN	
102	**TAROKO NATIONAL PARK**	P.172
	THAILAND	
103	**KHAO SOK NATIONAL PARK**	P.173
	TURKEY	
104	**GÖREME NATIONAL PARK**	P.176
	UNITED ARAB EMIRATES	
105	**WADI WURAYAH NATIONAL PARK**	P.177
	VIETNAM	
106	**PHONG NHA-KẺ BÀNG NATIONAL PARK**	P.178
107	**CAT BA NATIONAL PARK**	P.179

EUROPE

	ALBANIA	
108	**BUTRINT NATIONAL PARK**	P.182
	BELGIUM	
109	**HOGE KEMPEN NATIONAL PARK**	P.185
	BULGARIA	
110	**RILA NATIONAL PARK**	P.186
	CROATIA	
111	**KRKA NATIONAL PARK**	P.187
112	**PLITVICE LAKES NATIONAL PARK**	P.190
	FINLAND	
113	**HOSSA NATIONAL PARK**	P.191
114	**OULANKA NATIONAL PARK**	P.192
	GERMANY	
115	**SAXON SWITZERLAND NATIONAL PARK**	P.194

OVERVIEW

	GREECE	
116	**OLYMPUS NATIONAL PARK**	P.195
	GREENLAND	
117	**NORTHEAST GREENLAND NATIONAL PARK**	P.198
	ICELAND	
118	**VATNAJÖKULL NATIONAL PARK**	P.199
	IRELAND	
119	**CONNEMARA NATIONAL PARK**	P.203
	ITALY	
120	**DOLOMITI BELLUNESI NATIONAL PARK**	P.204
121	**GRAN PARADISO NATIONAL PARK**	P.206
122	**CINQUE TERRE NATIONAL PARK**	P.207
	LATVIA	
123	**GAUJA NATIONAL PARK**	P.210
	NORWAY	
124	**JOSTEDALSBREEN NATIONAL PARK**	P.211
125	**JOTUNHEIMEN NATIONAL PARK**	P.214
	POLAND	
126	**TATRA NATIONAL PARK**	P.215
	PORTUGAL	
127	**PENEDA-GERÊS NATIONAL PARK**	P.216
	ROMANIA	
128	**CHEILE NEREI-BEUSNITA NATIONAL PARK**	P.218
	SCOTLAND	
129	**CAIRNGORMS NATIONAL PARK**	P.219
	SLOVAKIA	
130	**SLOVAK KARST NATIONAL PARK**	P.220
	SLOVENIA	
131	**TRIGLAV NATIONAL PARK**	P.221

	SPAIN	
132	**ORDESA AND MONTE PERDIDO NATIONAL PARK**	P.222
	SWEDEN	
133	**ABISKO NATIONAL PARK**	P.223
	THE NETHERLANDS	
134	**NATIONAAL PARK DE BIESBOSCH**	P.224
135	**HET NATIONALE PARK DE HOGE VELUWE**	P.225
136	**DE MEINWEG NATIONAL PARK**	P.228
	UKRAINE	
137	**PODILSKI TOVTRY NATIONAL PARK**	P.229
	UNITED KINGDOM	
138	**DARTMOOR NATIONAL PARK**	P.230
	WALES	
139	**SNOWDONIA NATIONAL PARK**	P.231

OCEANIA

	AUSTRALIA	
140	**ULURU-KATA TJUTA NATIONAL PARK**	P.234
141	**GREAT BARRIER REEF MARINE PARK**	P.235
142	**CAPE RANGE NATIONAL PARK**	P.238
143	**KAKADU NATIONAL PARK**	P.239
144	**LEEUWIN-NATURALISTE NATIONAL PARK**	P.240
	FIJI	
145	**BOUMA NATIONAL PARK**	P.241
	NEW ZEALAND	
146	**TONGARIRO NATIONAL PARK**	P.243
147	**FIORDLAND NATIONAL PARK**	P.244
148	**AORAKI/MOUNT COOK NATIONAL PARK**	P.246
149	**ABEL TASMAN NATIONAL PARK**	P.247
150	**WHANGANUI NATIONAL PARK**	P.250

AFRICA — ALGERIA

01 GOURAYA NATIONAL PARK

TO VISIT
BEFORE YOU DIE
BECAUSE

This park and biosphere reserve boasts nice beaches, walking trails, and a wide array of animals – all off the beaten track.

Named for the nearby Gouraya Mountain, which stands more than 2,100 feet tall, Gouraya National Park is located where the cliffs and the coast meet in Algeria's Bejaia Province. Among the park's most notable inland inhabitants are the Barbary macaque, a species of monkey endemic to North Africa, and the endangered Algerian hedgehogs, wild cats, and jackals. On the coast, marine mammal sightings could include sperm whales, short-beaked common dolphins, bottlenose dolphins, harbor porpoises, and the endangered Mediterranean monk seal, which can occasionally be spotted along the rocky shores. Birdwatchers also have the opportunity to observe a plethora of avian species, including eagles, falcons, and migratory birds that pass through the area throughout the year. There are also 13 villages within the reserve, the population of which is primarily of Berber origin.

www.unesco.org/en/mab/gouraya

AFRICA BOTSWANA

02 CHOBE NATIONAL PARK

TO VISIT
BEFORE YOU DIE
BECAUSE

Botswana's first national park, this is one of the most biologically diverse conservation areas in Africa.

There are more elephants in Botswana's Chobe National Park than anywhere else on the planet – the park, known as "The Land of The Giants," was specifically formed for their protection. Today there are more than 120,000 Kalahari elephants in this preserve, with some herds numbering in the hundreds. But they're not the only animals in the conservation area: there are also large numbers of buffalos, wild dogs, cheetahs, lions, giraffes, zebras, wildebeest, hippos, hyenas, and more. Part of the reason there's such a high concentration of wildlife is the Chobe River, an important water source, particularly in the dry season. That vital waterway is one of the best ways for humans to witness this animal kingdom – various companies offer boat safaris, which offer a unique vantage for wildlife watching. In addition to river safaris, Chobe offers classic land-based game drives, allowing visitors to explore more of the park's diverse ecosystems, including floodplains, grasslands, and woodlands.

www.chobe.com

AFRICA

DEMOCRATIC REPUBLIC
OF THE CONGO

03 VIRUNGA NATIONAL PARK

TO VISIT BEFORE YOU DIE BECAUSE

It's the best place to see the endangered mountain gorilla, in one of the most biologically diverse areas on the planet.

Located in the eastern Democratic Republic of the Congo, this vast wilderness encompasses a staggering array of landscapes, from volcanic peaks and dense forests to savannahs and swamps. The park is best known for the mountain gorillas that inhabit it. It is estimated that around 350 of the world's total 1,050 mountain gorillas live here. Visitors can spend time with some of the habituated primates (our closest animal relatives) on guided treks. Other wildlife within the park includes lions, hippos, crocodiles, leopards, and buffalos, among others, as well as myriad bird species. Despite the natural and cultural riches, Virunga National Park faces numerous threats, including poaching, habitat destruction, and political instability. Conservation efforts are ongoing, led by dedicated park rangers and conservation groups, to protect the park's biodiversity and ensure the survival of its iconic species.

virunga.org

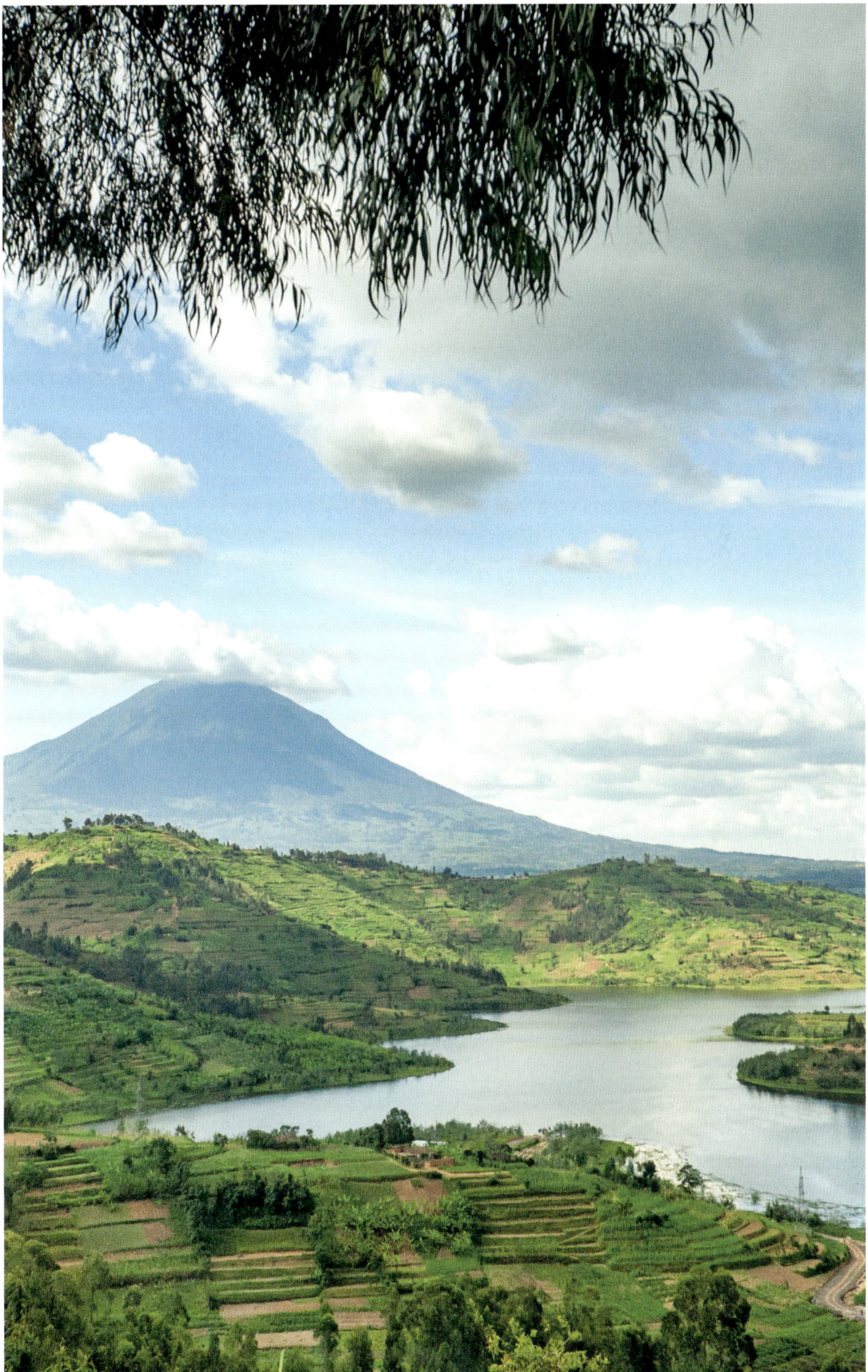

AFRICA   EGYPT

04   WADI EL GEMAL NATIONAL PARK

TO VISIT BEFORE YOU DIE BECAUSE

This unique desert and coastal wilderness includes both extraordinary ecosystems and numerous archaeological sites.

Spanning 2,880 square miles, roughly a third of which is marine space and two-thirds is land, Wadi El Gemal National Park is a protected area on Egypt's Red Sea coast. The name translates to "Valley of the Camel," which reflects the area's historical significance as a trading route for camel caravans. This park boasts a surprising array of animals, considering much of it is desert. The coastal area features more than 450 species of coral, 1,200 species of fish (including emperor angelfish, parrotfish, white-spotted puffer, and giant moray), and aquatic animals like sea turtles and spinner dolphins. Inland, you might spot the jerboa, the dorcas gazelle, the Barbary sheep, and the Nubian ibex. Beyond an exciting destination for animal lovers, Wadi El Gemal is also an intriguing spot for history buffs. The park is home to several archaeological sites dating back thousands of years, including rock art, ancient settlements, and remnants of Roman and Byzantine civilizations.

AFRICA EGYPT

05 WHITE DESERT NATIONAL PARK

TO VISIT
BEFORE YOU DIE
BECAUSE

The otherworldly white rock formations of this red sand desert, which is home to fennec foxes and sand cats, are absolutely stunning.

Famed for its large, Seussically strange white rock formations, created by erosion caused by wind and sand, White Desert, also known as Sahara el Beyda, is one of the most unique and dreamy national parks in Africa. The rocks (made of either white calcium, quartz, chalk, or limestone) have whimsical shapes resembling everything from mushrooms to flying saucers to animals, depending on your imagination. In some places, the smaller formations simply look like piles of snow in the desert. The figures are made more surreal at sunrise and sunset, when they're painted shades of pink and orange. The park serves as a refuge for many animals, including Barbary sheep, jackals, rhim and dorcas gazelles, red and fennec foxes, and the sand cat. Visitors can embark on guided tours or self-drive expeditions, getting out of the car to hike amongst the more interesting formations and to discover the park's many hidden treasures, from ancient fossil beds to archaeological sites.

national-parks.org/egypt/white-desert

AFRICA   ETHIOPIA

06 SIMIEN MOUNTAINS NATIONAL PARK

TO VISIT
BEFORE YOU DIE
BECAUSE

It is the largest national park in Ethiopia and home to species seen nowhere else, including the Gelada baboon and Abyssinian catbird.

Though chances are you've never heard of Simien Mountains National Park, the protected land is incredibly important for its biodiversity – in fact, it was one of the very first sites to be named a World Heritage Site by Unesco, for that reason. One of the most iconic inhabitants of the Simien Mountains is the Gelada baboon, often referred to as the "bleeding-heart monkey" due to the distinctive red patch of skin on its chest. The caracal, the Walia ibex, and the Ethiopian wolf are also endemic to the region (and the latter two of which are endangered). For birdwatchers, the Simien Mountains offer a veritable paradise, with over 180 avian species recorded within the park. Among the most sought-after sightings is the lammergeier, or bearded vulture, a formidable bird of prey known for its impressive wingspan and scavenging habits. Other notable species include the Abyssinian catbird, the white-collared pigeon, and the rare Ethiopian bush crow.

simienpark.org

AFRICA KENYA

07 HELL'S GATE NATIONAL PARK

татTO VISIT BEFORE YOU DIE BECAUSE

This remarkable quarter of the Great Rift Valley inspired the setting of classic Disney movie *The Lion King*.

The park's name, "Hell's Gate," evokes images of a fiery inferno, but in reality it's a serene destination – it was named after a break in the cliffs, once a tributary of a prehistoric lake. Still, two extinct volcanoes, Olkaria and Hobley's, are located in the park, as are some hot springs that are warm enough to cause burns. A quick day trip from Nairobi, it's an easy safari destination, where visitors can see a variety of wildlife, including buffalos, giraffes, lions, baboons, hyena, zebra, and various antelope species, amidst savannah grasslands and acacia woodlands. It's also a prime destination for birdwatchers, with more than 100 species of feathered creatures having been recorded within the park's boundaries, including eagles, flamingos, and the rare lammergeier vulture. Visitors can hike, bike, and motorcycle within the park, a true rarity in African safari destinations. There are also a handful of campsites and several lodges for visitors to spend the night within the protected area.

www.kws.go.ke/content/hells-gate-national-park

AFRICA　　　　　KENYA

08　LAKE NAKURU NATIONAL PARK

TO VISIT BEFORE YOU DIE BECAUSE

Some of the largest population of flamingos in the world are found in this birdwatchers' paradise, which is also home to rhinos and giraffes.

Lake Nakuru is perhaps most famous for its prolific birdlife. There are over 450 bird species, both endemic and migratory, to the park. However, it's really one bird that draws the most visitors: the flamingo. Lake Nakuru is a haven for millions of flamingos, whose vibrant pink plumage creates a stunning spectacle against the backdrop of the shimmering lake. Visitors can witness these elegant birds wading through the shallows in search of their favorite food, algae-rich crustaceans. Beyond its avian inhabitants, Lake Nakuru National Park is also home to a diverse mammalian population. The park is renowned for its protection of endangered black and white rhinoceros (more than 100 live within the borders), offering visitors a rare opportunity to observe these majestic creatures up close in their natural habitat. Additionally, herds of buffalos, zebras, giraffes, and antelopes roam the grasslands while elusive predators such as lions, leopards, and hyenas lurk in the shadows, adding an element of excitement to every safari adventure.

www.kws.go.ke/lake-nakuru-national-park

AFRICA MADAGASCAR

09 ISALO NATIONAL PARK

TO VISIT
BEFORE YOU DIE
BECAUSE

A visit to this park is the opportunity to fill up your memory card with photos of lemurs and otherworldly sandstone rock formations.

While Madagascar as a whole is known for its diverse flora and fauna (scientists estimate that roughly 90 percent of the plants and 85 percent of the animals on this island nation are endemic), what sets Isalo National Park apart is its scenery – though don't worry animal lovers, there are still more than a dozen lemur species, a slew of reptiles and amphibians, and at least 80 bird species for your viewing pleasure. Located in the south of the country, the park is dominated by a sandstone massif that over the course of thousands of years has eroded into a mosaic of colorful canyons, gorges, pinnacles, and plateaus. As such, hiking is one of the biggest draws of the park. Piscine Naturelle Trail is a favorite option of travelers, as it terminates at a crystalline watering hole. Another option is the Namaza Circuit, which takes hikers through a landscape of forests, grasslands, and rocky outcrops. It's a requirement to have a local guide to accompany you through the park, which can be hired at the entrance, if you aren't already on an organized tour.

parcs-madagascar.com/parcs/isalo

AFRICA NAMIBIA

10 SKELETON COAST NATIONAL PARK

TO VISIT
BEFORE YOU DIE
BECAUSE

Dotted with hundreds of shipwrecks, this park is a hauntingly beautiful and desolate wilderness located along Namibia's rugged coastline.

Stretching over 6,500 square miles, this unique park is renowned for its stark landscapes, dramatic sand dunes, and shipwrecks scattered along its ocean-battered shores. The name "Skeleton Coast" derives from the whale and seal bones that once littered the beaches, remnants of the whaling industry that once thrived here, as well as the wreckage of countless vessels that have met their fate on the treacherous shores over centuries due to thick fog, stormy weather, and unpredictable currents. Even today, it's possible to see the remains of more than 1,000 ships. Despite its forbidding name and harsh conditions, Skeleton Coast National Park is home to a surprising array of wildlife, including desert-adapted elephants, rhinos, jackals, giraffes, ostriches, kudus, zebras, lions, gemsbok, and brown hyenas – mostly towards the northeastern corner of the park. On the shores you will find plenty of seals, with dolphins and whales putting in occasional but spectacular appearances in the ocean.

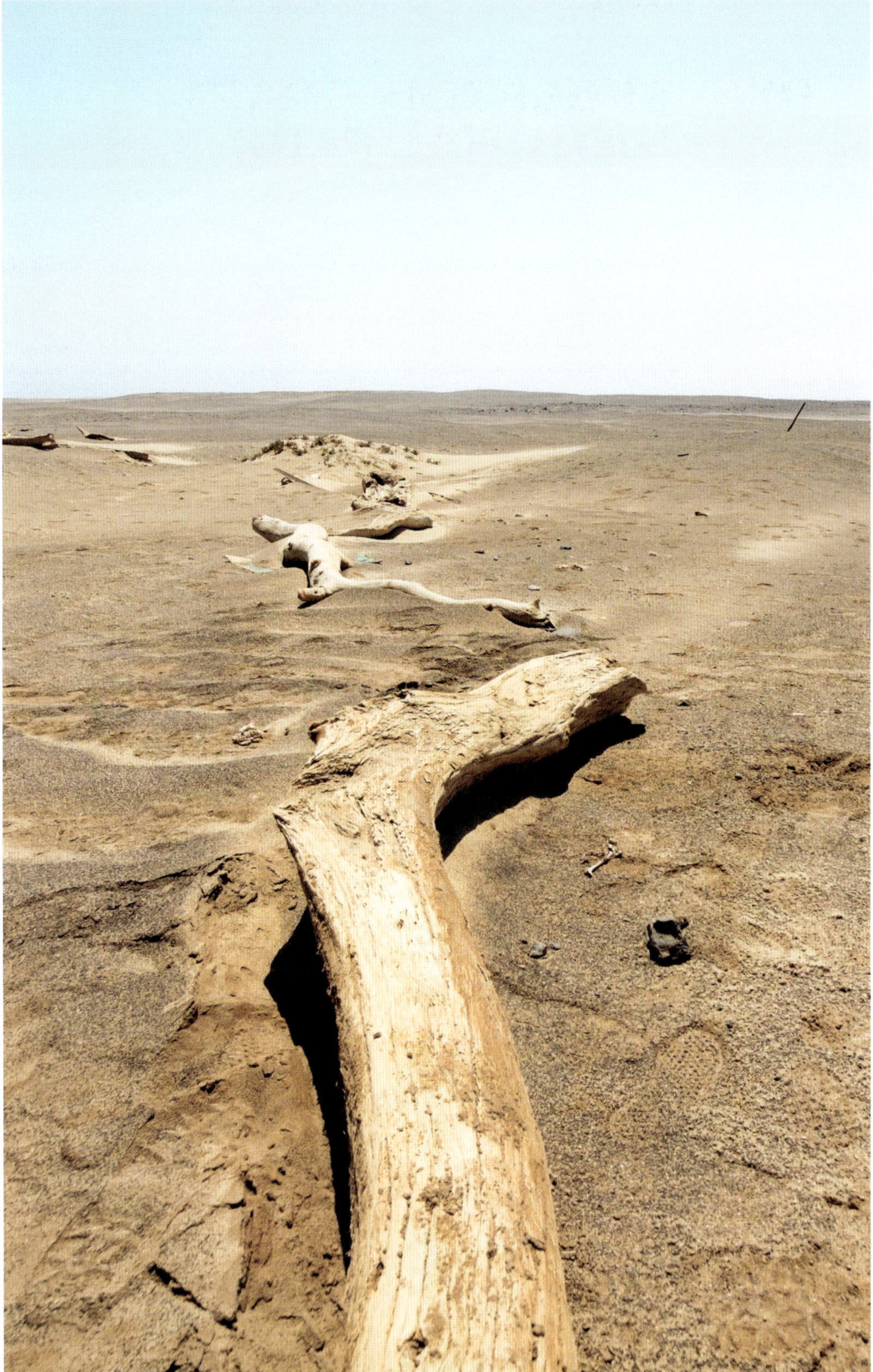

AFRICA NAMIBIA

11 ETOSHA NATIONAL PARK

TO VISIT BEFORE YOU DIE BECAUSE

Boasting a flat, open salt pan so big it is visible from space, this park is one of the premier wildlife viewing destinations on the African continent.

One of the largest national parks in Africa, the Etosha conservation area sprawls approximately 8,600 square miles in northwestern Namibia. The name "Etosha" translates to the "Great White Place" in the language of the Ovambo people, a moniker it derives from a colossal salt flat that accounts for roughly a quarter of the park's territory. The pan itself, though seemingly desolate most of the year, transforms during the rainy season into a shallow, ephemeral lake, attracting flocks of flamingos and other water birds in breathtaking numbers. Given how sparse the vegetation in Etosha is, it's easy to spot wildlife, including iconic African megafauna such as lions, giraffes, cheetahs, black and white rhinos, Burchell's zebras, blue wildebeest, and elephants (the tallest in Africa). The park's waterholes serve as vital lifelines during the dry season, drawing in animals by the thousands and offering unparalleled opportunities for wildlife viewing and photography.

AFRICA　　　　　　　　NAMIBIA

12　NAMIB-NAUKLUFT NATIONAL PARK

TO VISIT BEFORE YOU DIE BECAUSE

Set in the world's oldest desert, this vast park is known for its immense red sand dunes and dramatic Sossusvlei and Deadvlei areas.

Namib-Naukluft National Park, the largest national park on the continent, sprawls across the arid landscapes of Namibia's southwestern region. Its biggest draw is the Namib Desert, one of the oldest and driest deserts on Earth, characterized by towering red sand dunes (some of the tallest in the world) stretching endlessly. Salt and clay pans are also common visitor attractions – the most popular include Sossusvlei, Hiddenvlei, and Deadvlei. Here, visitors may encounter elusive desert-adapted species such as oryx, springbok, snakes, geckos, hyenas, gemsboks, springboks, bat-eared foxes, and even the rare desert elephant. You can also spot thousands-year-old welwitschia plants and some of the local little Five, including the Namaqua chameleon, sand-dwelling spider, and Peringuey's adder. For those seeking adventure, Namib-Naukluft National Park offers a wealth of outdoor activities, from hiking and 4x4 desert drives to going on scenic flightseeing tours or hot air balloon expeditions over the vast dunes. The area is also a Dark Sky Reserve, so stargazing is a must.

visitnamibia.com.na/namib-naukluft-park-2

AFRICA REPUBLIC OF THE CONGO

13 ODZALA-KOKOUA NATIONAL PARK

TO VISIT BEFORE YOU DIE BECAUSE

At the heart of the Congo Basin and the second-largest tropical rainforest on Earth, this remote park is one of the planet's last true wildernesses.

Part of the larger Congo Basin, the 90-year-old Odzala-Kokoua National Park in the central African country of the Congo is an incredibly valuable habitat, not just to the 10,000 species of animals that live there, including the critically endangered western lowland gorillas and the forest elephants, but to the world. Considered "the lungs of Africa," this rainforest is the largest carbon sink in the world, absorbing more carbon than the Amazon. It is estimated that only a few hundred people visit Odzala-Kokoua National Park a year, making it one of Africa's last true wilderness frontiers – much of the park hasn't been explored by outsiders. One of the most exciting ways to experience the park is on a gorilla trek. A primatologiost named Magda Bermejo was the first to habituate the western lowland gorillas in the 90s – work she still does today from her base at Kamba, the only safari lodge operator in the park. Other popular ways for visitors to explore the park are on guided canoe or bai safaris, as a network of rivers and wetlands wind their way through the dense forest and are home to an abundance of animal species including crocodiles, buffalos, hippos, fish, and birds.

14 RÉUNION NATIONAL PARK

TO VISIT BEFORE YOU DIE BECAUSE

This tiny French-speaking Indian Ocean island has more than 200 microclimates and more than 800 endemic species.

Roughly 42 percent of the island of Réunion, an overseas department of France in the western Indian Ocean between Madagascar and Mauritius, is made up of Réunion National Park. At the heart of the park is Piton de la Fournaise, one of the world's most active volcanoes, which continually shapes the island's dramatic topography. Its lunar landscapes provide a stark contrast to the verdant forests that blanket much of the park's lower elevations. There are myriad adventures to be had within the park. With its rugged terrain and numerous waterfalls, it is a prime destination for canyoning and guided tours allow visitors to rappel down cascades, glide down natural rockslides, and swim through crystal-clear pools, providing an exhilarating way to experience the park's natural beauty. There are also myriad trails, though the most popular include Grand Bassin, La Roch Merveilleuse, and Sentier Du Bras Mapou. From majestic land turtles to shy green geckos or the many species of lizards, you'll mostly encounter reptiles while hiking on Réunion Island.

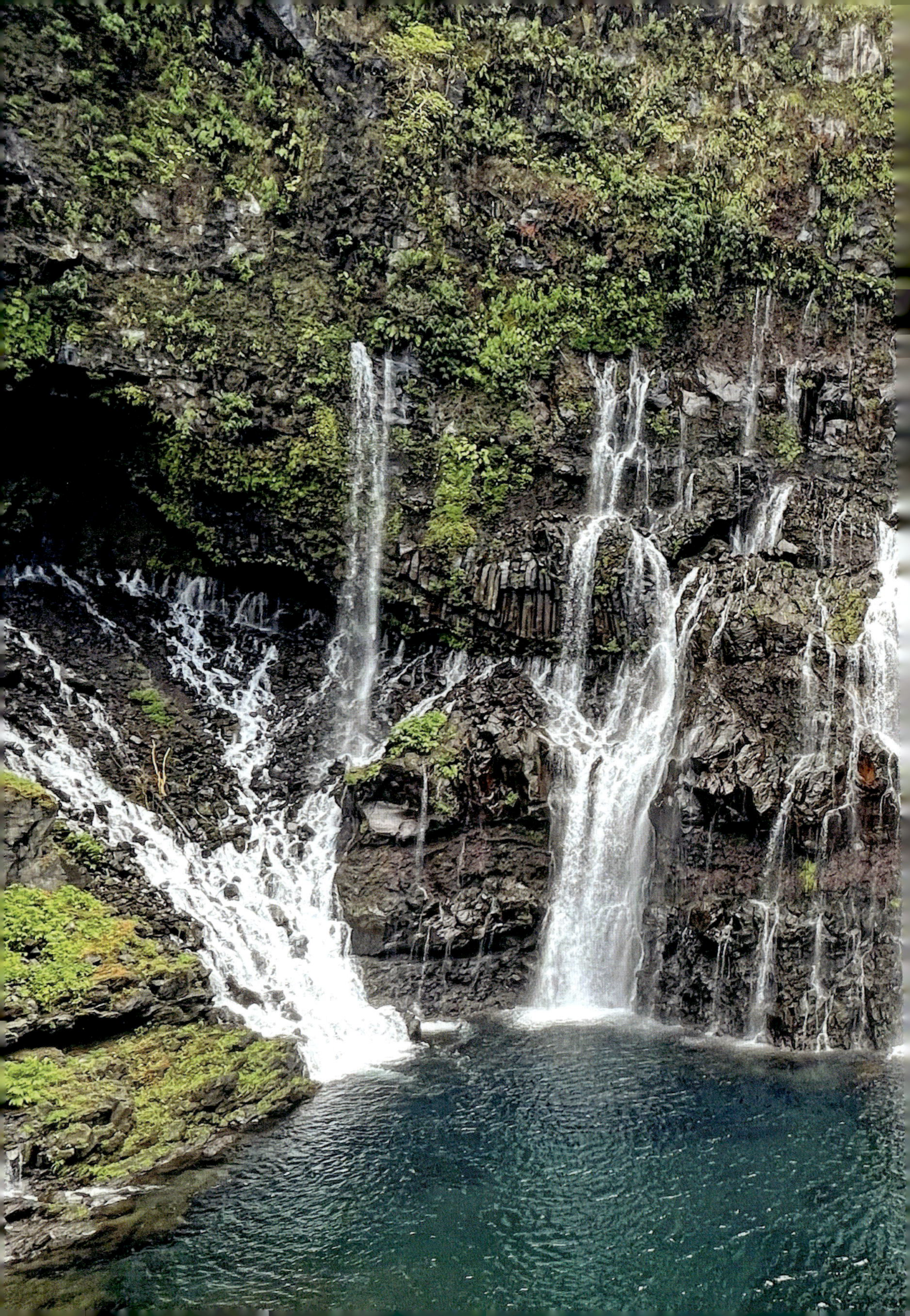

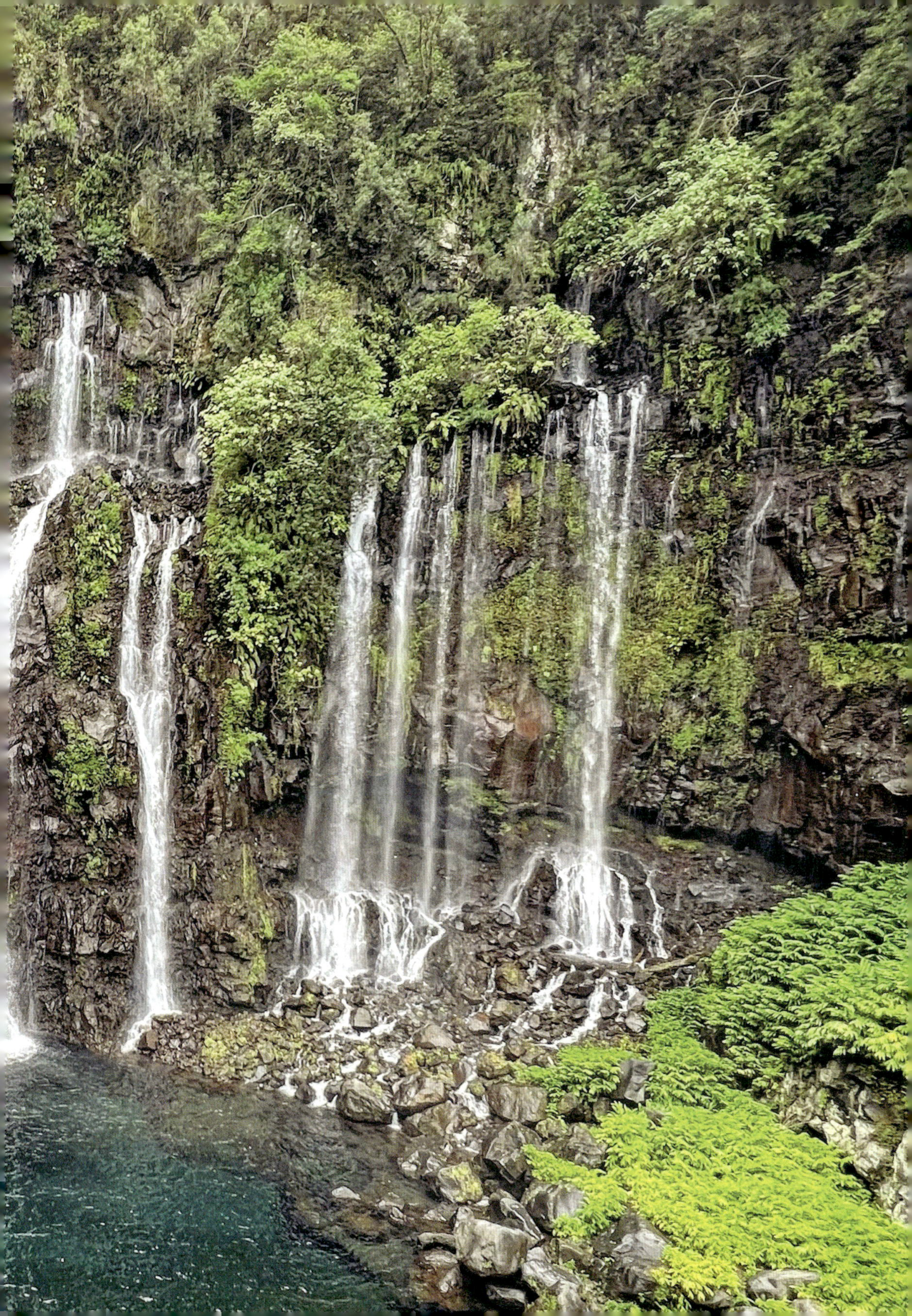

AFRICA　　　　RWANDA

15　VOLCANOES NATIONAL PARK

TO VISIT BEFORE YOU DIE BECAUSE

High on the Virunga Massif, this central African national park is where famed primatologist Diane Fossey lived and worked.

Chances are, if you've heard of Volcanoes National Park, it is because you've read the book *Gorillas in the Mist* or seen the film adaption by the same name, which chronicles the life of Diane Fossey, an American primatologist and conservationist who devoted her life to studying and protecting the mountain gorillas of Rwanda. Today, visitors to the park can see the descendants of the gorillas that she worked with on a gorilla trekking expedition. Led by experienced guides and trackers, travelers can come face-to-face with these great apes in their natural habitat, providing an intimate and unforgettable wildlife encounter. There are roughly a dozen mountain gorilla families (and about 400 individual primates) found throughout the park. Before or after your hike, it's worth stopping by the new campus of the Diane Fossey Gorilla Fund. As of 2022, there is now an excellent museum dedicated to teaching visitors more about animals and the continued conservation efforts.

visitrwanda.com/destinations/volcanoes-national-park

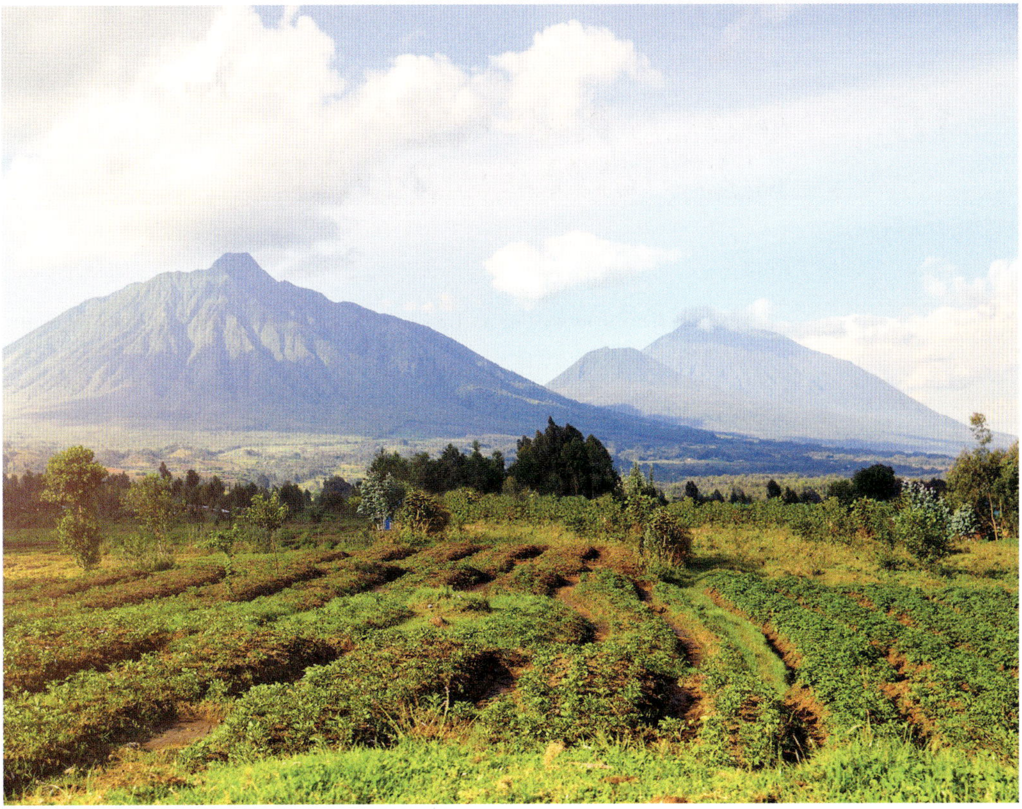

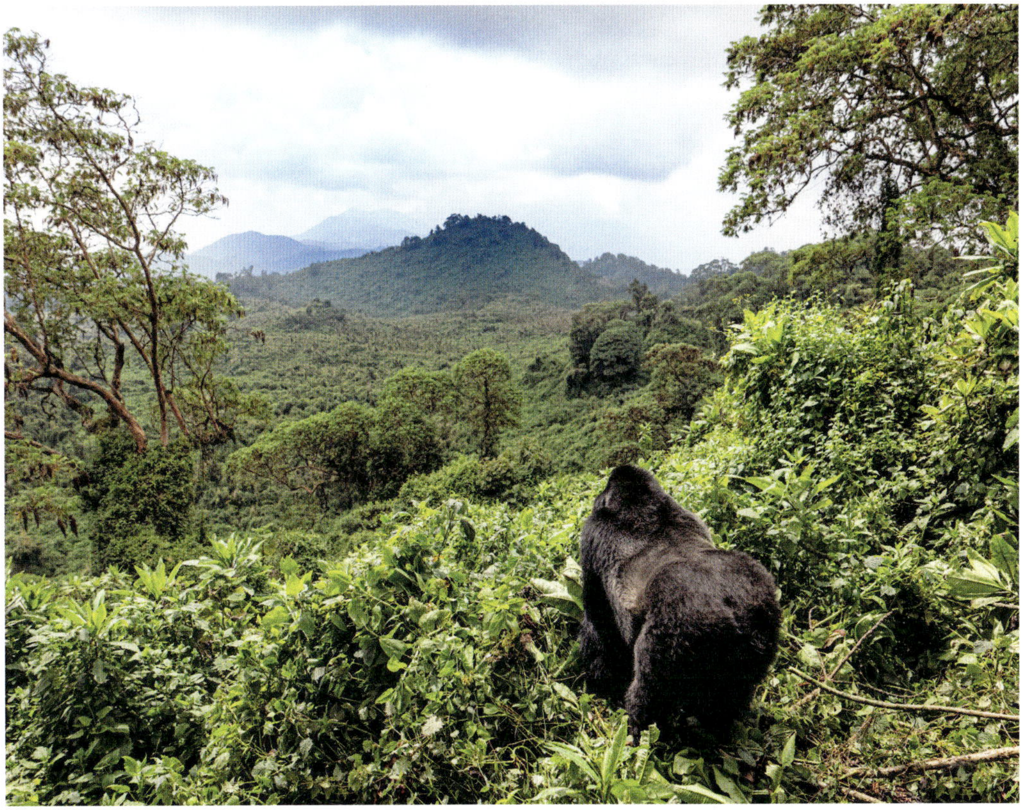

AFRICA SENEGAL

16 NATIONAL PARK LANGUE DE BARBARIE

TO VISIT BEFORE YOU DIE BECAUSE

Every winter, this park becomes a vitally important stopover and breeding ground for thousands of migratory birds from Europe.

Langue de Barbarie National Park, which translates to "Tongue of the Barbarian," gets its name due to its location, an elongated sandy peninsula which juts out into the Atlantic Ocean. However, its shape is not what this Unesco Biosphere Reserve is best known for. In reality, the park's most prominent feature is its vital role as a sanctuary for a wide array of avian species, particularly from November to April, making it a paradise for birdwatchers and ornithologists. It serves as an essential stopover and breeding ground for numerous migratory birds from Europe, including pelicans, flamingos, terns, spoonbills, herons, and beyond. More than 160 species of birds can be found throughout the park, meaning it plays a crucial role in conservation efforts and in protecting vulnerable ecosystems. The site also hosts numerous reptile species, crabs, monkeys, and five species of threatened marine turtles. If you choose to come to the park independently, know that hiring a guide from the park office is mandatory.

rsis.ramsar.org/ris/2467

AFRICA   SOUTH AFRICA

17 KRUGER NATIONAL PARK

TO VISIT BEFORE YOU DIE BECAUSE

It's the largest national park in Africa, the oldest in South Africa, and a testament to conservation and biodiversity.

One of Africa's premier wildlife reserves, Kruger National Park boasts an incredibly diverse ecosystem comprising savannahs, woodlands, and forests, which provide a habitat for an astounding array of wildlife. Here, you can encounter Africa's iconic "Big Five" – lions, elephants, buffalos, leopards, and rhinoceroses – along with a plethora of other species, including giraffes, zebras, cheetahs, hyenas, and countless bird species. Visitors to Kruger National Park are treated to unforgettable safari experiences, whether on guided game drives, walking safaris, or self-drive adventures along well-maintained roads. The park's numerous rest camps, lodges, and private reserves offer accommodation options ranging from luxurious to rustic, ensuring there's something for every type of traveler. It's also worth noting that the park is an exciting destination for archaeological enthusiasts – there is evidence that prehistoric men lived in the area over half a million years ago.

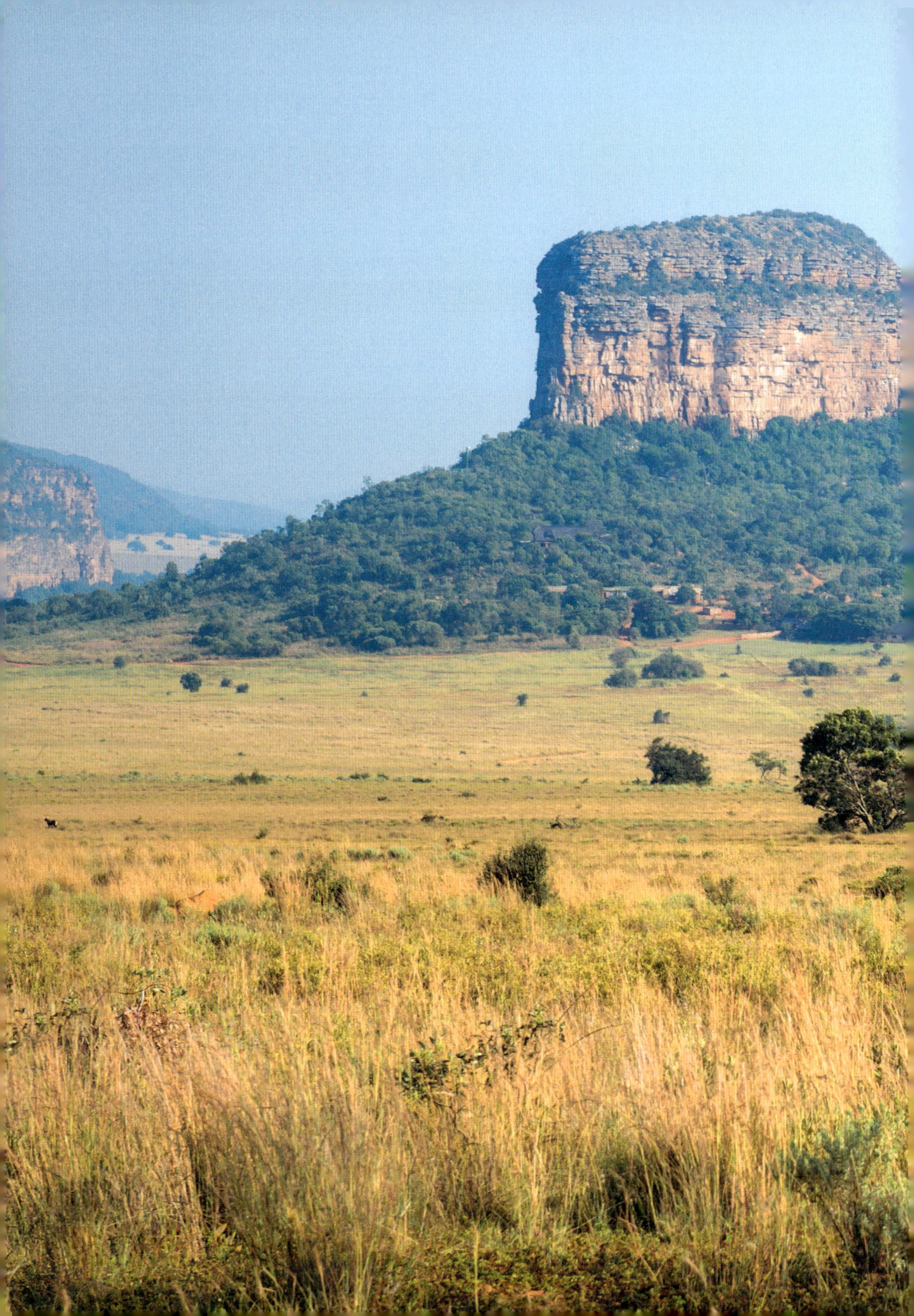

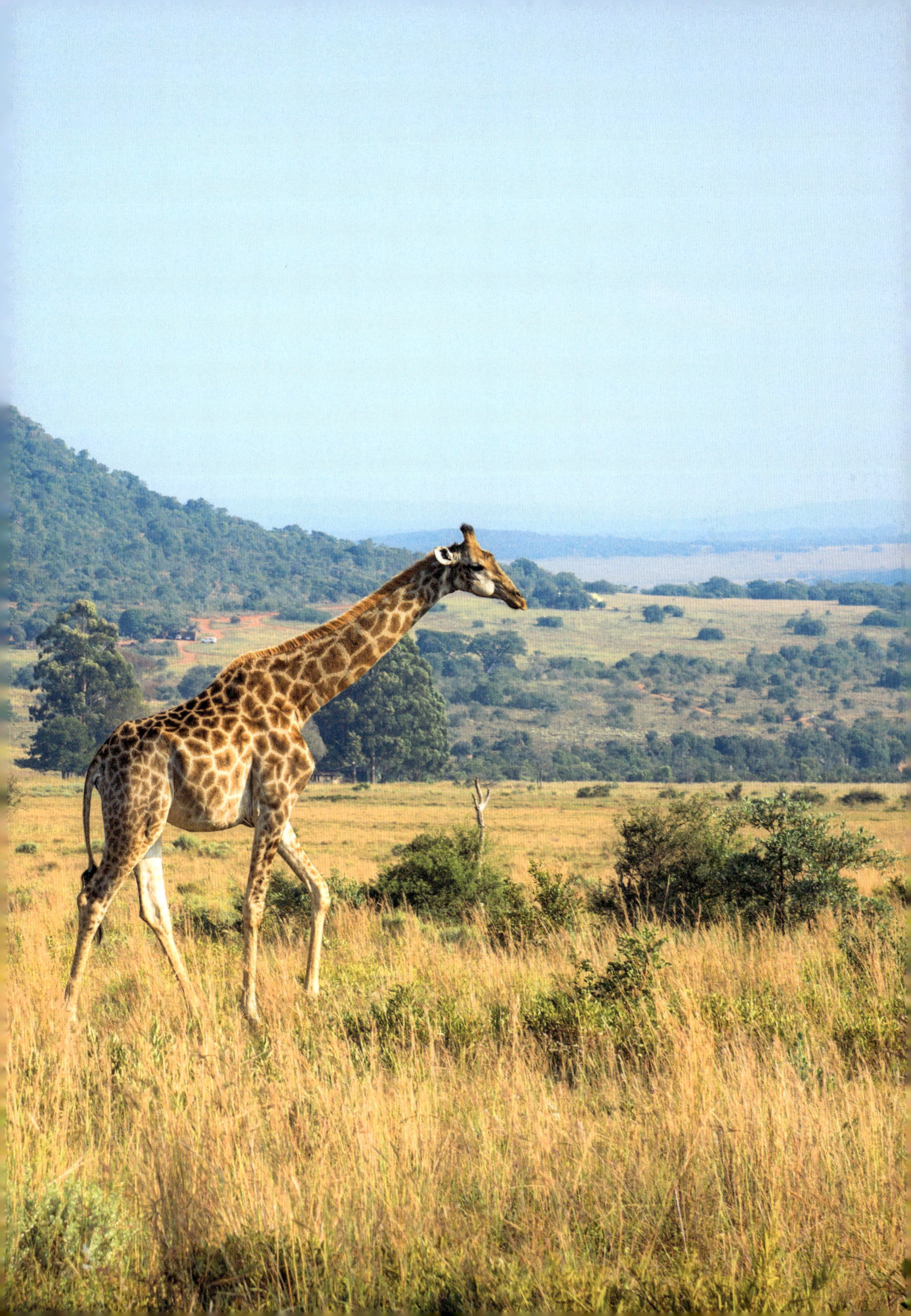

AFRICA SOUTH AFRICA

18 TABLE MOUNTAIN NATIONAL PARK

TO VISIT BEFORE YOU DIE BECAUSE

Rich in wildlife and endemic plants, yet close to the city, this park protects one of the most culturally significant mountains in Africa.

With its distinctive flat-topped summit, often shrouded in a blanket of clouds known locally as the "tablecloth," Table Mountain is the geological marvel that gives this South African national park its name. Getting atop the mountain is one of the most popular activities here – there are oodles of trails, each with varying degrees of difficulty, to reach the summit. There is also an aerial cableway that can whisk guests to the top in under five minutes. Despite being so close to Cape Town, one of the largest cities in the country, there's an incredible array of wildlife, including Cape mountain zebra, chacma baboon, rock hyrax, Cape fox, and Cape clawless otter. There are also many avian residents, ranging from raptors to tiny sun birds. Beyond its outdoor activities and animals, the park also holds cultural significance to the local Khoukhoi and San communities, who consider it a spiritual place and a symbol of their heritage. The park stretches all the way south to the spectacular Chapman's Peak Drive, unmissable Boulders Penguin Colony, and unspoiled fynbos landscapes of Cape Point and Cape of Good Hope.

AFRICA　　　　　　　TANZANIA

19　SERENGETI NATIONAL PARK

TO VISIT BEFORE YOU DIE BECAUSE

Renowned for its breathtaking landscapes and abundant wildlife, the Serengeti hosts the world's most spectacular land animal migration.

Located in northern Tanzania, Serengeti National Park is one of the most revered wildlife sanctuaries in the world. It is home to an extraordinary array of wildlife, making it one of the best places in Africa for safari experiences. The park is renowned for its large populations of big cats, including lions, leopards, and cheetahs, which roam the plains in search of prey. Other predators such as hyenas and wild dogs are also commonly sighted. The park's herbivores include vast herds of wildebeest, zebras, gazelles, and buffalos, which undertake the annual Great Migration in search of fresh grazing grounds. This natural spectacle, considered one of the Seven Natural Wonders of Africa, sees millions of animals trekking across the Serengeti, accompanied by predators and scavengers. Visitors to Serengeti National Park have a range of activities to choose from to explore this remarkable wilderness. The most popular activity is game driving, either in safari vehicles or on guided walking safaris.

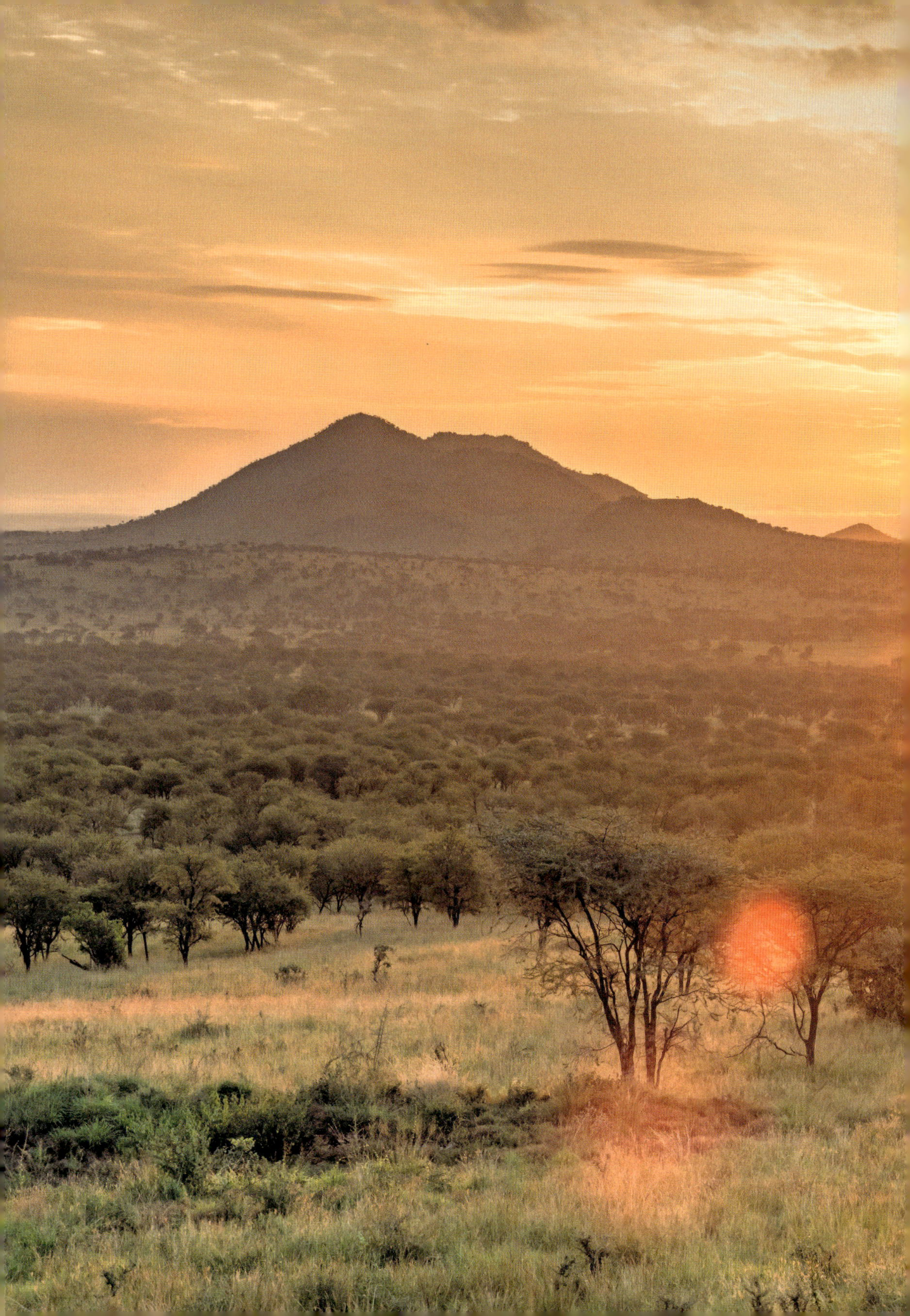

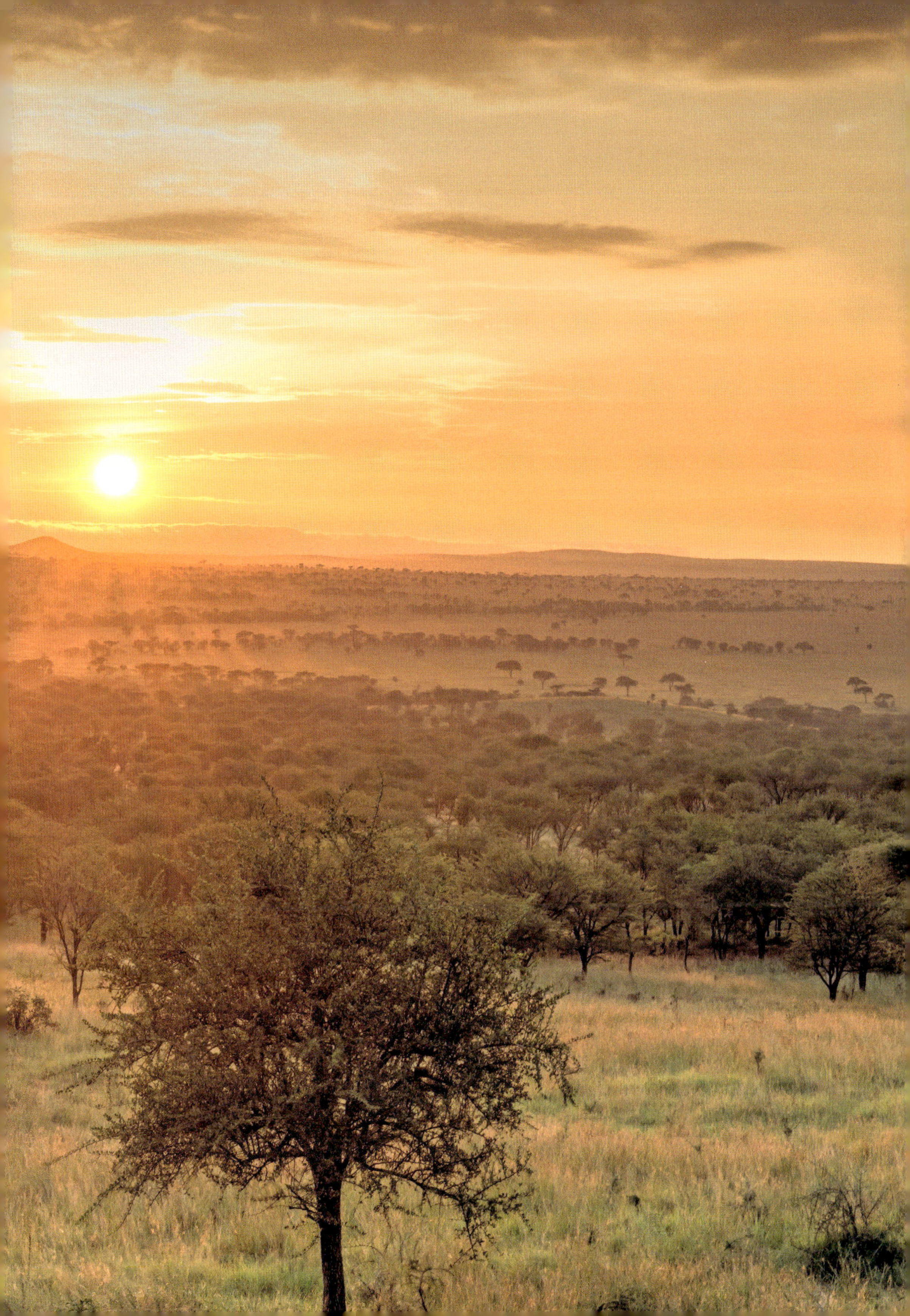

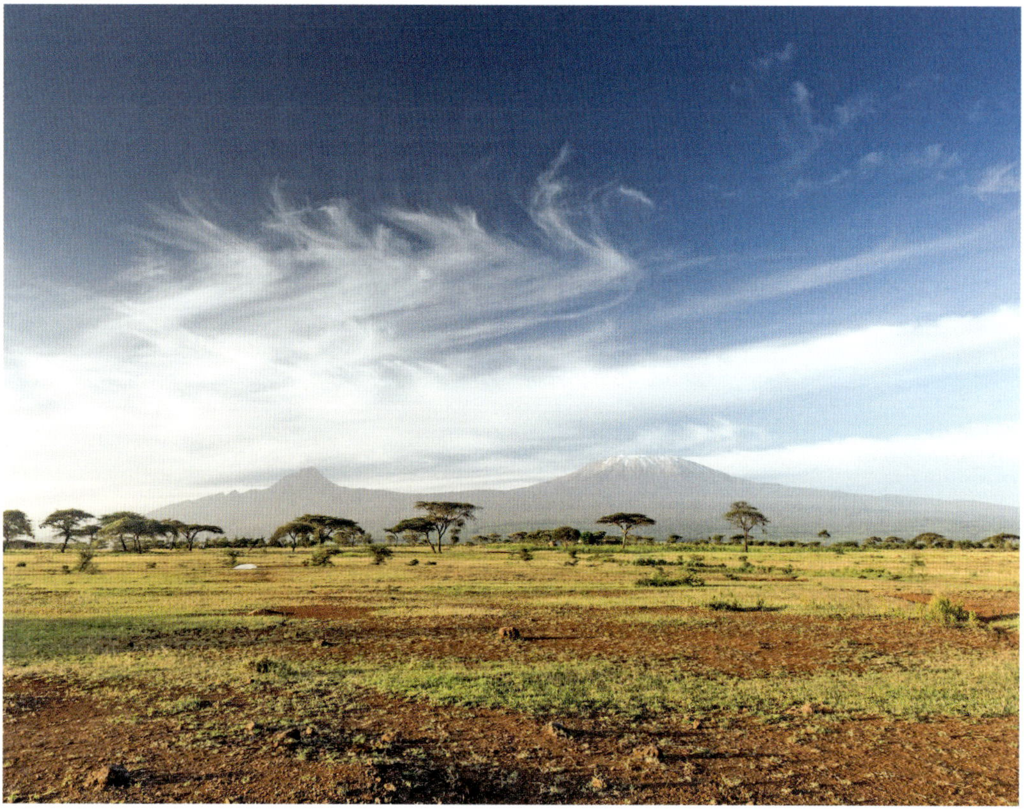

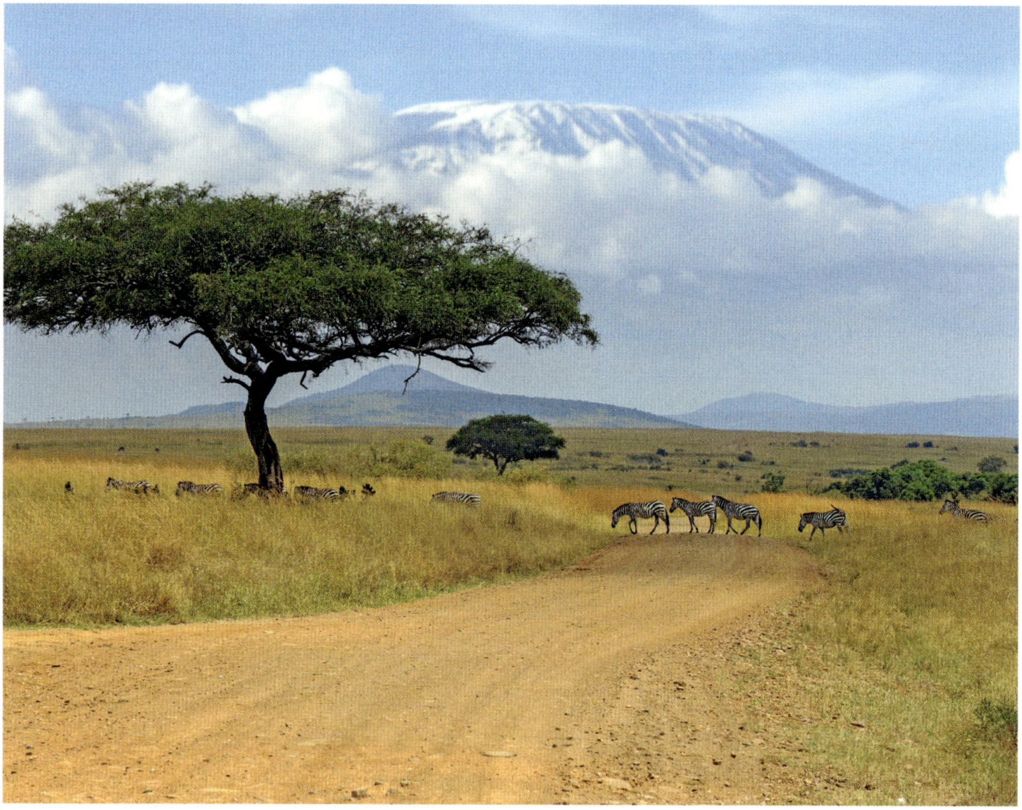

AFRICA   TANZANIA

20 KILIMANJARO NATIONAL PARK

TO VISIT BEFORE YOU DIE BECAUSE

This national park is home to the majestic Mount Kilimanjaro, the "roof of Africa" and the largest free-standing mountain in the world.

Tanzania's Kilimanjaro National Park is a Unesco World Heritage Site and home to one of Africa's most iconic natural landmarks: Mount Kilimanjaro, Africa's highest peak and the tallest free-standing mountain in the world. Standing at 19,341 feet above sea level, the famous volcanic massif is a challenging yet rewarding climb for adventurers from around the globe and offers several trekking routes, each with its own distinct scenery and challenges, catering to both novice hikers and experienced mountaineers. The snow-capped peak is surrounded by vast savannahs that support a huge number of animals, making the park a popular safari spot in addition to a hiking destination. Some of those creatures include elephants, Cape buffalos, giraffes, leopards, hyenas, baboons, mongooses, honey badgers, and bush babies. There are also some critically endangered black rhinos. Visitors might also glimpse some of the myriad birds in the park, ranging from vultures and buzzards to African pygmy kingfishers and malachite sunbirds.

www.tanzaniaparks.go.tz

AFRICA TANZANIA

21 NGORONGORO NATIONAL PARK

TO VISIT
BEFORE YOU DIE
BECAUSE

Home to the Maasai people, this extinct volcanic crater boasts the densest population of lions in the world, and is a stunning location to spot the "Big Five."

Millions of years ago a volcano collapsed, forming what is now the Ngorongoro Crater. This natural amphitheater stretches over 100 square miles and has a thickly forested rim that towers up to 2,000 feet above the floor of the caldera. Within its confines, a vast and fertile grassland supports a dense population of wildlife, making it one of the best places in Africa to spot the "Big Five" (lions, elephants, buffalos, leopards, and rhinoceroses), as well as zebras, warthogs, wildebeests, Grant's and Thomson's gazelles, and more. Beyond the crater, the park extends into the highlands, with rolling hills, volcanic peaks, and vast plains. The Maasai, a group of nomadic pastoralists and their livestock, not only inhabit the area but also contribute significantly to the cultural richness of the park. Their presence adds a unique dimension to the park's tapestry, inviting visitors to appreciate their way of life.

www.tanzaniaparks.go.tz

AFRICA   UGANDA

22 BWINDI IMPENETRABLE NATIONAL PARK

TO VISIT BEFORE YOU DIE BECAUSE

You can meet our closest living animal relatives, the mountain gorillas, in their natural setting, a mist-covered rainforest.

Spread over a series of steep ridges and valleys in southwestern Uganda, this national park and Unesco World Heritage Site stands as one of the last remaining strongholds of the critically endangered mountain gorillas. Home to approximately half of the world's remaining population of these great apes, the park offers a rare opportunity for intimate encounters with mountain gorillas. Each day, experienced guides lead small groups on strenuous treks through the tangled undergrowth of the forest, through thickets of ferns and bamboo, in search of one of the family groups of gorillas. Once found, visitors are allowed one hour with the gorillas, where they can observe their natural behaviors (which are surprisingly humanlike) up close. Beyond the iconic gorillas, Bwindi (which means "darkness" in the local language) is also the habitat for chimpanzees, elephants, buffalos, bush pigs, side-striped jackals, civets, and more than 350 species of birds, many of which are endemic to the region.

ugandawildlife.org/national-parks/bwindi-impenetrable-national-park

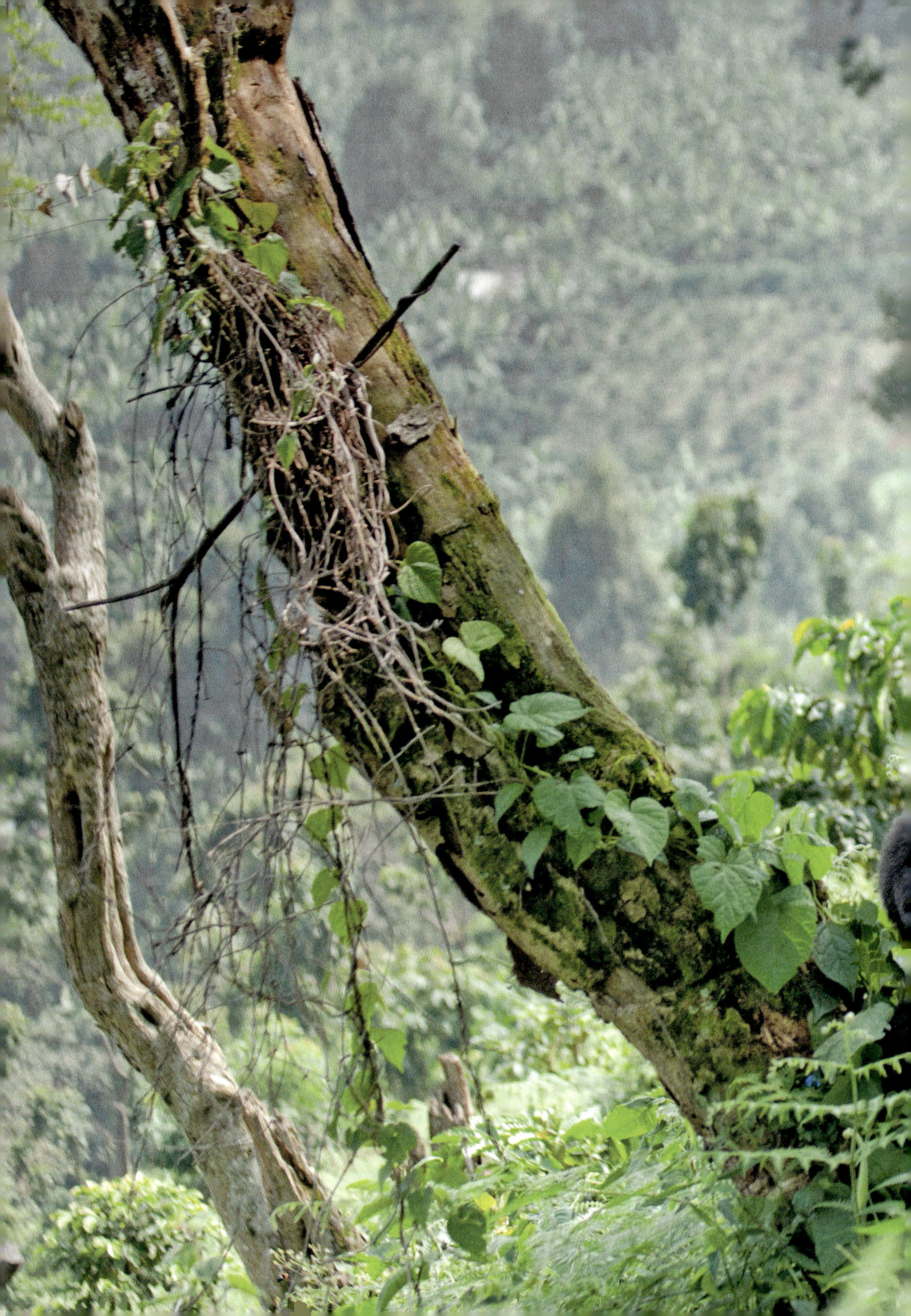

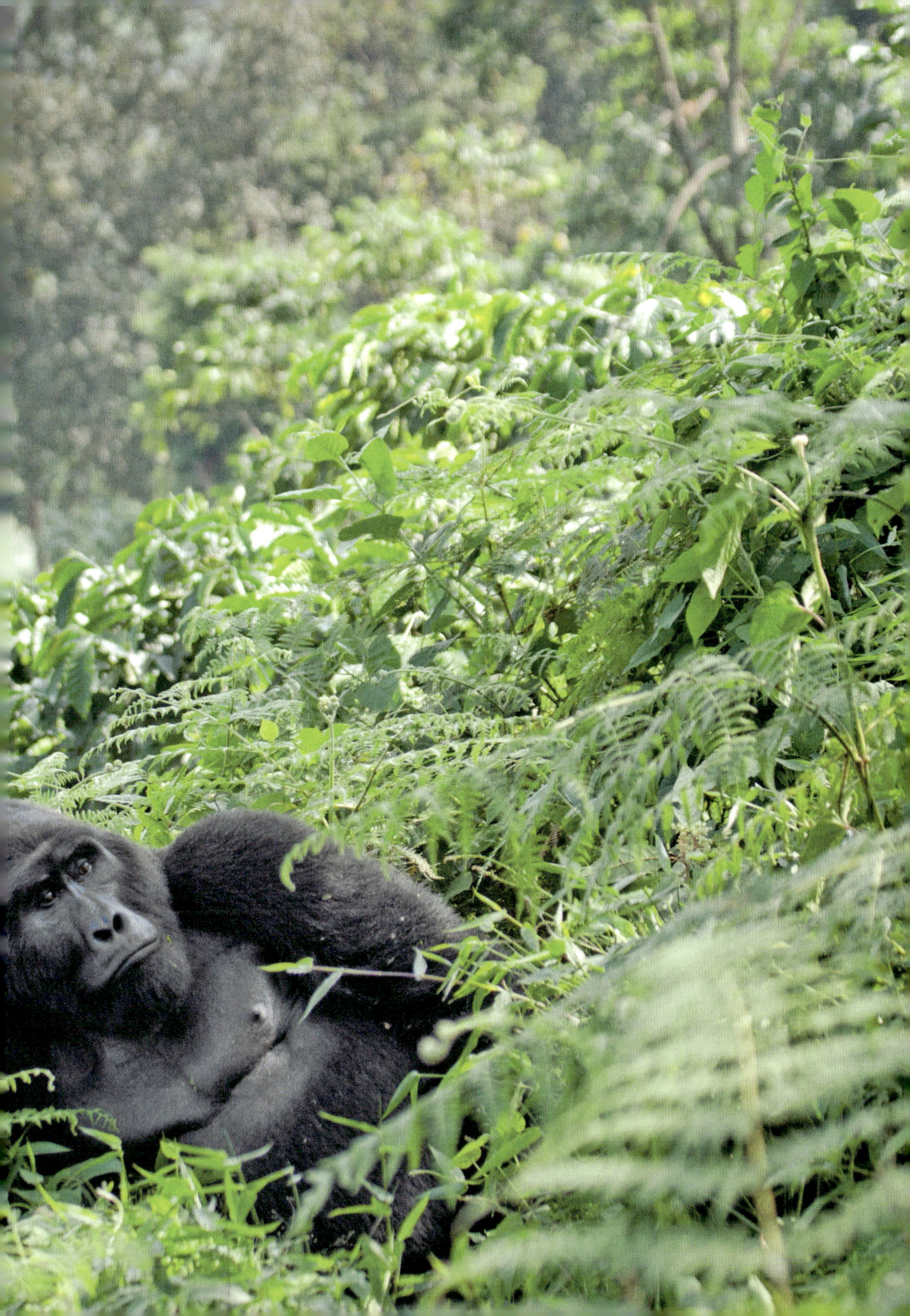

23 SOUTH LUANGWA NATIONAL PARK

TO VISIT BEFORE YOU DIE BECAUSE

The birthplace of walking safaris, this national park is one of the greatest wildlife sanctuaries in the world.

The Luangwa River, from which this park takes is name, is the most intact major river system in Africa and supports a staggering array of wildlife, including elephants, lions, zebras, hippos, baboons, buffalos, crocodiles, and Thornicroft's giraffes (which are found nowhere else). There are also more than 400 species of birds, as well as magnificent tamarind trees and some incredible baoba specimens. Though perhaps what gave rise to the popularity of the region is the fact that it is where walking safaris originated. In the 1950s, a conservationist named Norman Carr pioneered the practice, which is still done today. While the game drives here are incredible, there's something special about walking across the savannah, looking for signs of nearby wildlife on a more microscopic level, such as paw prints in the red dirt, bent twigs and grass, and the remains of a recent meal. And there is arguably nothing more thrilling than peering through the brush to find the leopard you've been tracking all morning.

www.zambiatourism.com/destinations/national-parks/south-luangwa-national-park

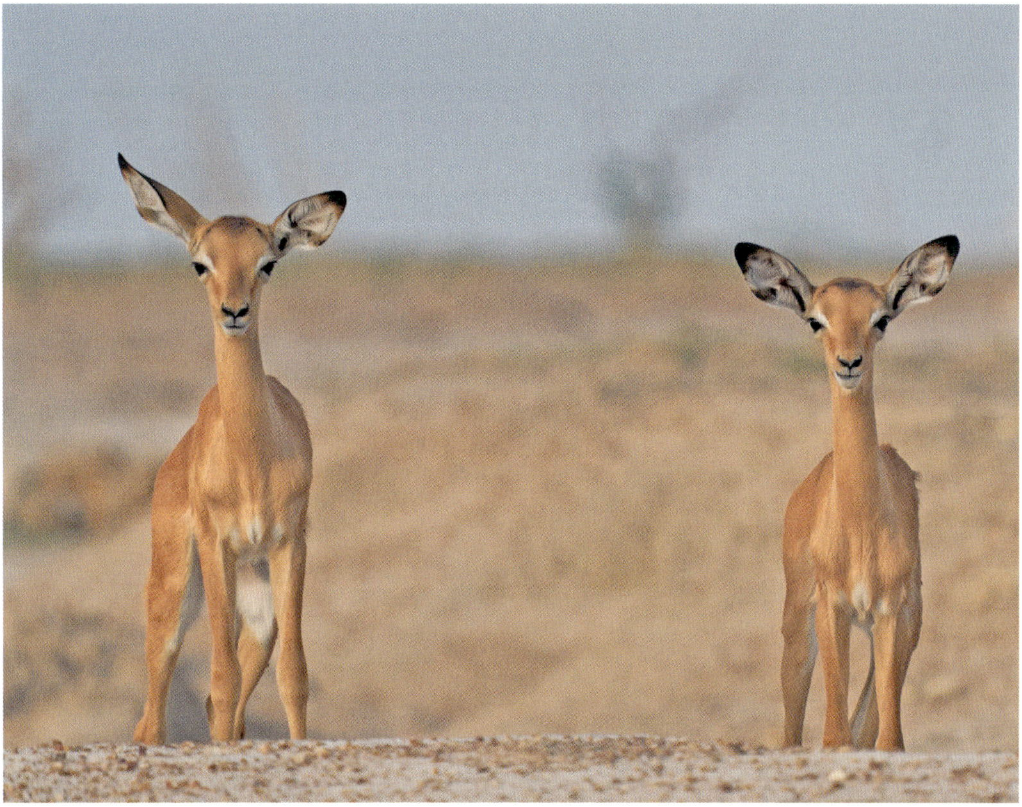

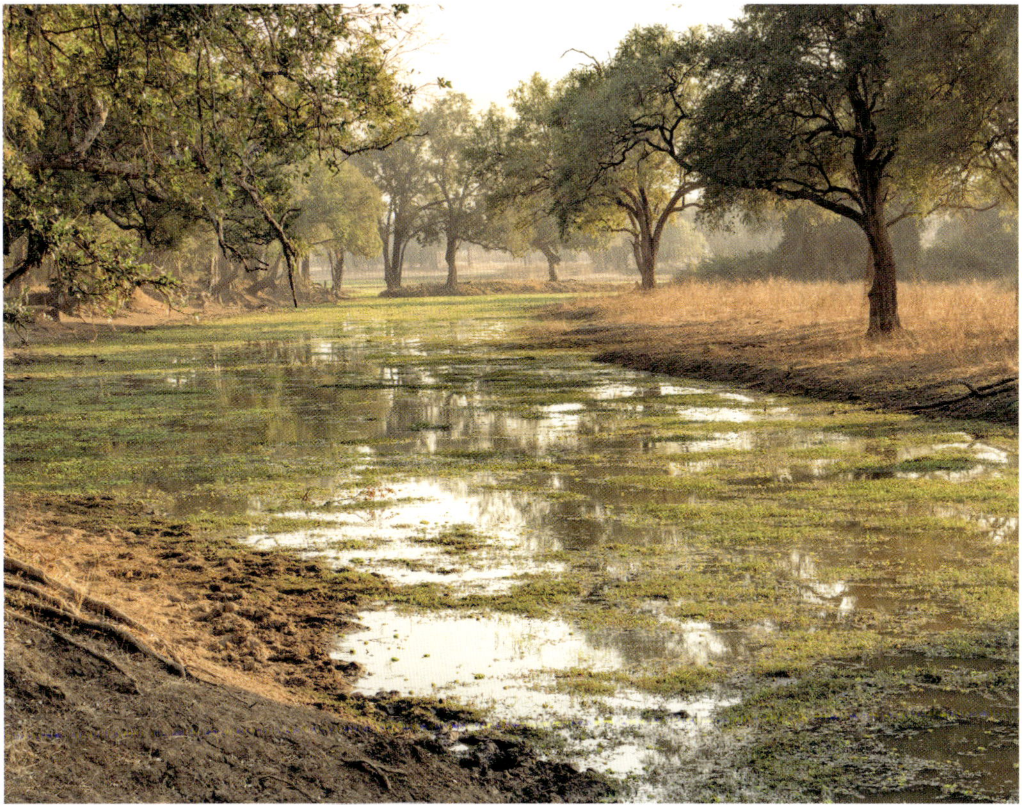

AFRICA · ZIMBABWE

24 HWANGE NATIONAL PARK

TO VISIT BEFORE YOU DIE BECAUSE

An incredible number of elephants dwell within this former game reserve turned national park in the west of the country.

Formerly known as Wankie Game Reserve, Hwange National Park is the largest natural reserve in all of Zimbabwe at more than 5,600 square miles. The park encompasses a variety of ecosystems, including vast grasslands, woodlands, and mopane forests, interspersed with natural water sources such as pans and seasonal streams. These diverse habitats support a remarkable array of flora and fauna. For instance, the park features a large population of elephants, estimated to be around 40,000, making it one of the densest concentrations of these majestic animals in Africa. Other prominent species in the park include lions, leopards, cheetahs, African wild dogs, buffalos, giraffes, zebras, and a wide variety of antelope species. There are also more than 400 species of birds. Visitors to Hwange National Park can enjoy a range of activities, including game drives, guided walks, and birdwatching safaris. The park offers several camps and lodges, providing accommodation options that cater to various preferences and budgets, from luxury tented camps to more rustic bush camps.

zimbabwetourism.net/portfolios/hwange-national-park

AFRICA　　　　　　　　　ZIMBABWE

25　VICTORIA FALLS NATIONAL PARK

TO VISIT BEFORE YOU DIE BECAUSE

To see the wonder that is Victoria Falls, the largest waterfall in the world, and embark on thrilling adventures such as whitewater rafting.

The centerpiece of this Zimbabwean national park is undoubtably Victoria Falls, known locally as Mosi-oa-Tonya, meaning "The Smoke That Thunders." This majestic cascade plunges over 100 meters into the Batoka Gorge, sending up a mist that can be seen from miles away. The sheer force of the falls is a marvel, which can be witnessed from several viewpoints along the park's network of trails, offering unique perspectives and photo opportunities. One of the most dramatic spots is the westernmost panorama known as Cataract View. Danger View is perhaps the most exciting, albeit spine-tingling spot to witness the curtain of water. This Unesco World Heritage Site and national park encompasses not only the falls and a stretch of the Zambezi River, but also the surrounding rainforest ecosystem, which is full of sunbirds and monkeys. Adrenaline-seekers can embark on thrilling adventures such as whitewater rafting, bungee jumping, zip-lining, going for a scenic helicopter ride over the falls, and more.

zimbabwetourism.net/portfolios/victoria-falls

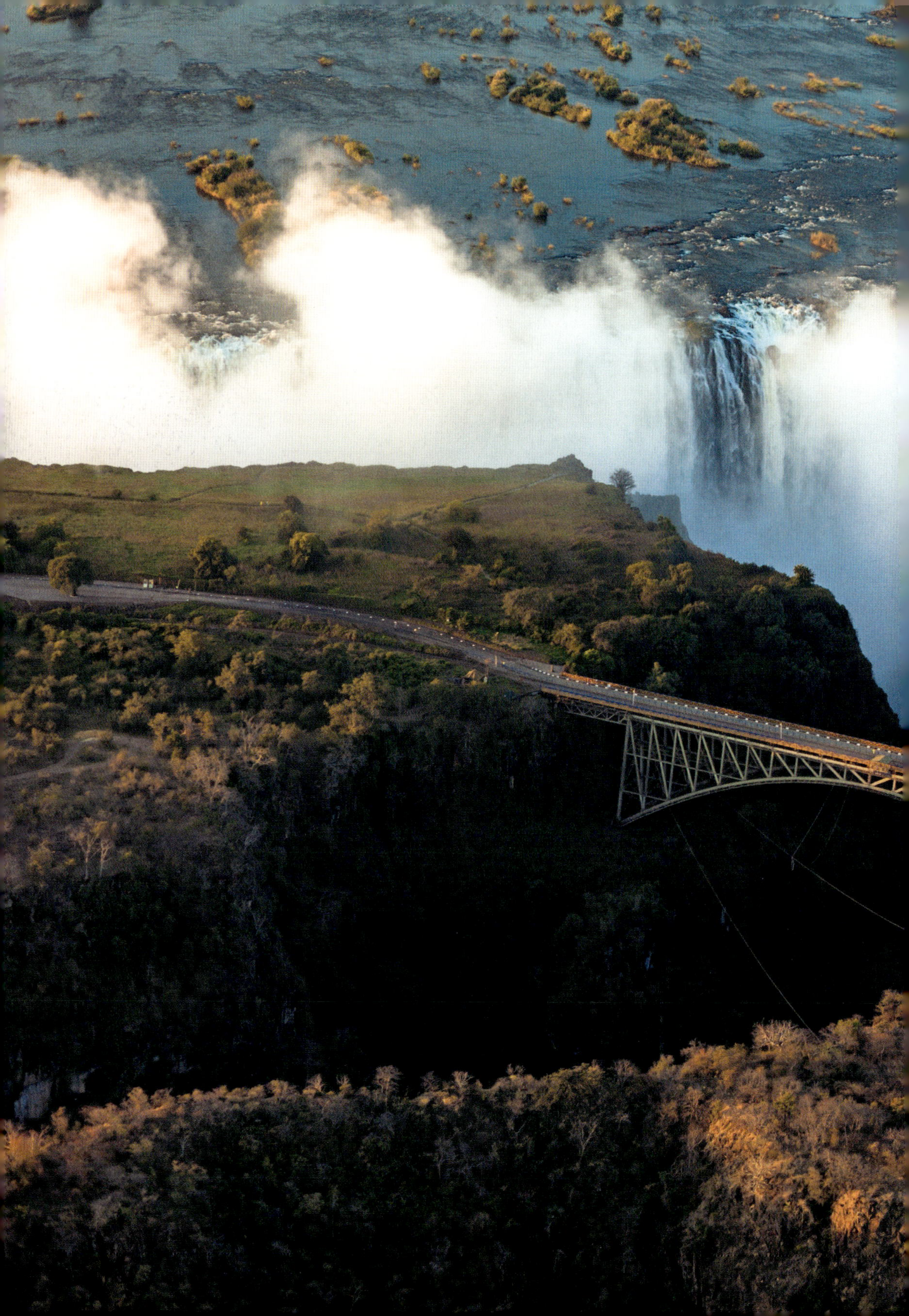

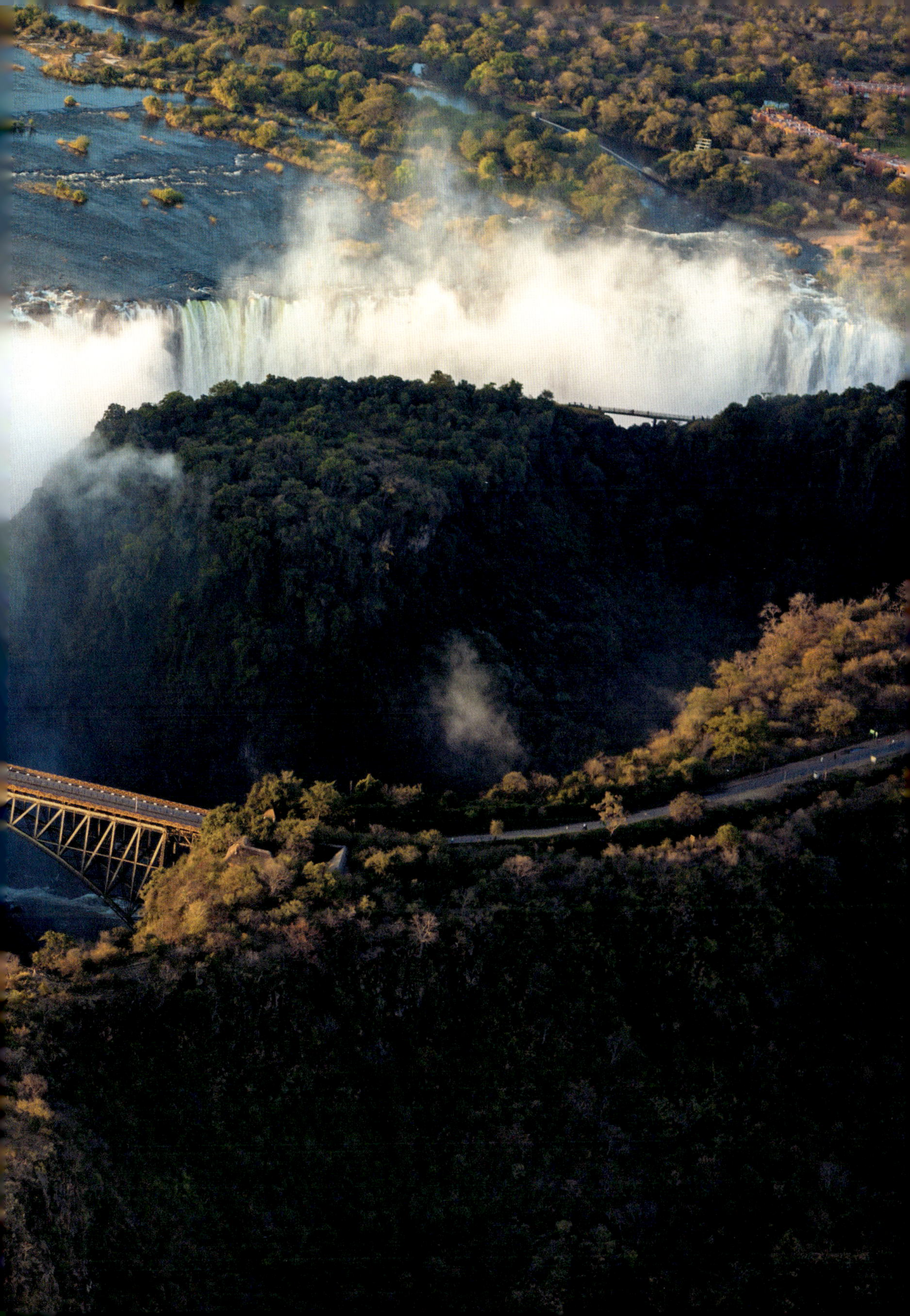

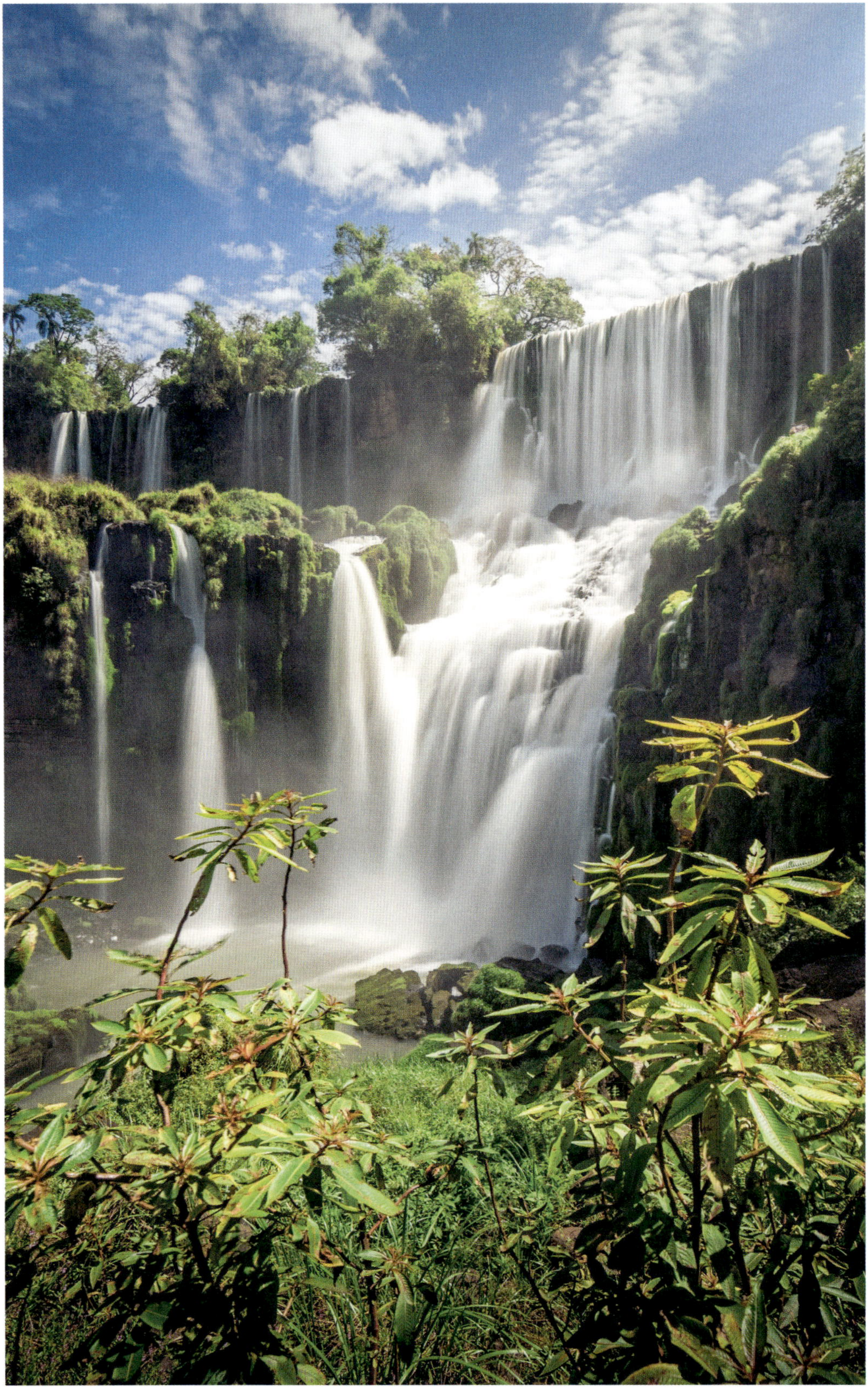

THE AMERICAS ARGENTINA/BRAZIL

26 IGUAZÚ NATIONAL PARK

TO VISIT
BEFORE YOU DIE
BECAUSE

This slice of South America is home to the world's largest waterfall and lush subtropical jungle teeming with life.

Found on the border of Argentina and Brazil, Iguazú Falls is the largest waterfall in the world, with 275 separate cascades spread out over nearly two miles along the Iguazú River. The massive falls plunge dramatically into the gorge below, creating a mesmerizing spectacle of mist, rainbows, and roaring water. One of the most iconic viewpoints within the park is the awe-inspiring Devil's Throat ("Garganta del Diablo"), where the Iguazú River plunges over a precipice, sending torrents of water crashing into the chasm below. The sheer power and beauty of this natural show leave a lasting impression on all who behold it (Eleanor Roosevelt famously said "My poor Niagara" after witnessing the cascade). That constant spray creates an extremely humid microclimate, which allows over 2,000 species of plants to grow within the park. The territory, set within the Alto Paraná Atlantic Forests ecoregion, is also home to animals such as the giant anteater, howler monkey, jaguar, tapir, and ocelot.

THE AMERICAS — ARGENTINA

27 TIERRA DEL FUEGO NATIONAL PARK

TO VISIT BEFORE YOU DIE BECAUSE

Jump aboard the End of the World Train to discover this remote region combining jaw-dropping marine, forest, and mountain environments.

Translating to "Land of Fire," Tierra del Fuego earned its name from Spanish explorers who observed myriad bonfires lit by the local Indigenous people as they sailed past this landscape at the southern tip of South America. Here, where the Andes Mountains meet the icy waters of the Southern Ocean, is an area renowned for its islets, glaciers, lakes, snow-capped mountains, and semi-deciduous forests. It is where animals such as the Andean fox, the guanaco, southern river otter, seals, austral parakeet, Patagonian woodpecker, Magellanic oystercatcher, kelp goose, albatross, petrels and penguins call home. Hiking along the well-marked trails and kayaking in the many waterways are popular pursuits in Tierra del Fuego National Park. Also of note is the End of the World Train, a historic railway that travels through the scenic valleys and beech forests of Argentina. The southern terminus of the Pan-American Highway is also located within the park.

turismoushuaia.com/zonas/parque-nacional

THE AMERICAS ARGENTINA

28 LOS GLACIARES NATIONAL PARK

TO VISIT
BEFORE YOU DIE
BECAUSE

Head to the Austral Andes to see Perito Moreno Glacier, one of the few ice masses in the world that is still growing.

On the southern reaches of Argentina's Patagonia region is a reserve that pays tribute to the raw power and beauty of nature: Los Glaciares National Park. At the heart of the park lies the Southern Patagonian Ice Field, one of the largest ice fields outside of Antarctica, from which numerous glaciers flow into the surrounding valleys. Among these, the most famous is the colossal Perito Moreno Glacier, a mesmerizing expanse of ice that stretches over 19 miles long and towers up to 240 feet above the surface of Lake Argentino, where it terminates. While most other glaciers in the world are retreating and shrinking, Perito Moreno Glacier stands out because it's continuously growing – each day, it advances an average of five feet. Beyond the glaciers, Los Glaciares National Park boasts diverse ecosystems, including rugged mountains, expansive forests, and crystal-clear lakes. The towering peaks of the Andes Mountains provide a stunning backdrop for outdoor enthusiasts, offering hiking, mountaineering, and wildlife viewing opportunities.

www.argentina.travel/en/actividades/los-glaciares-national-park

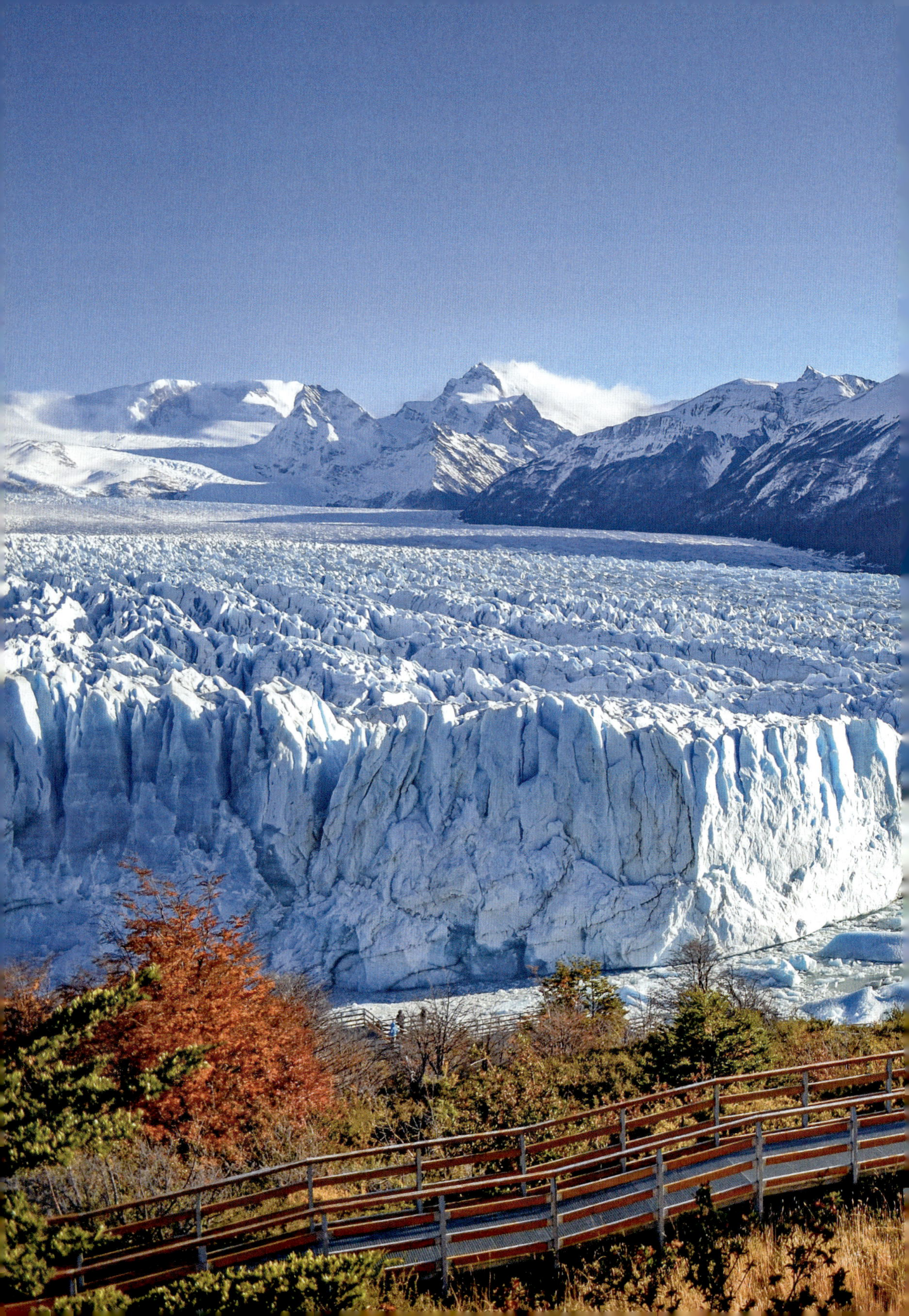

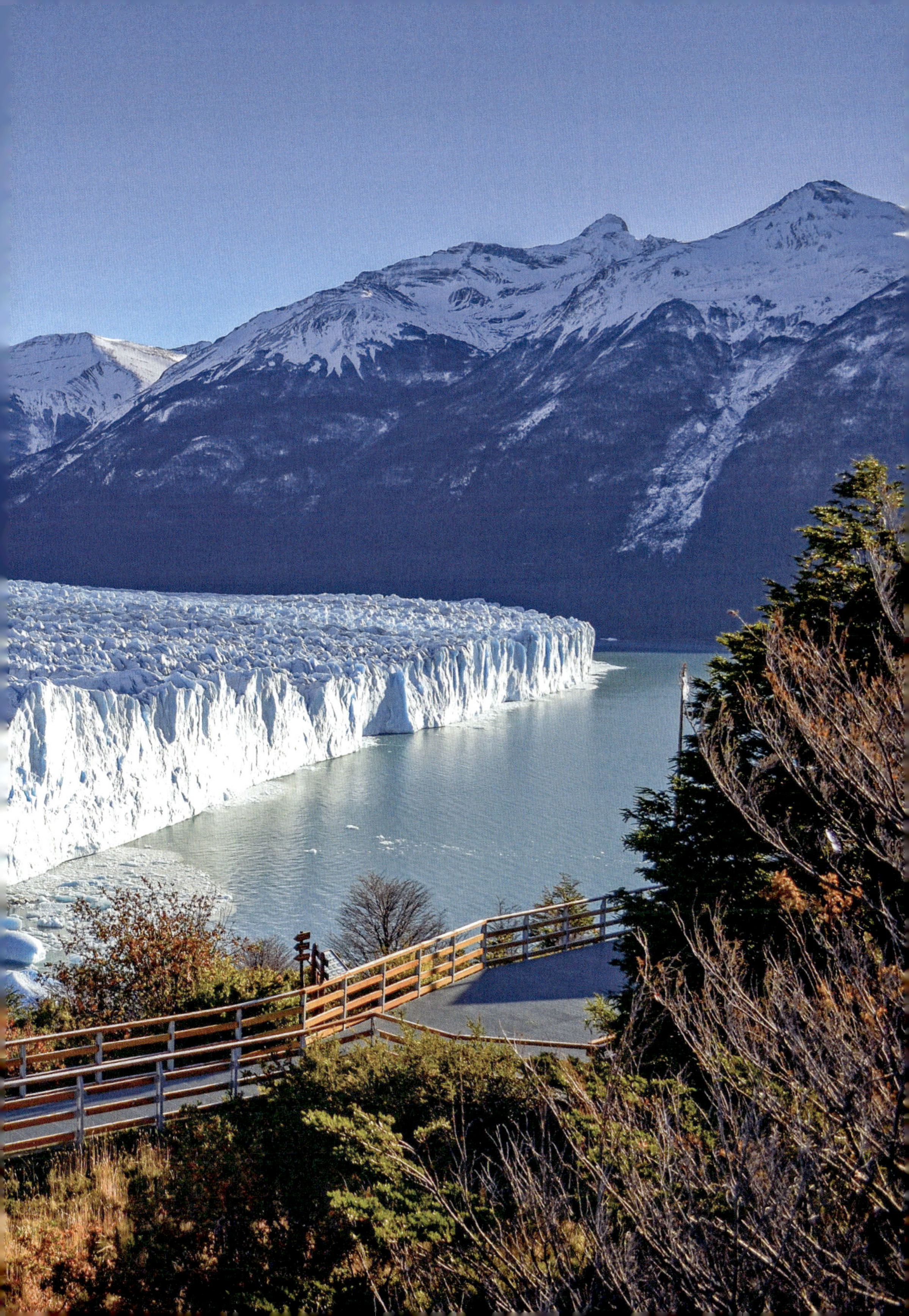

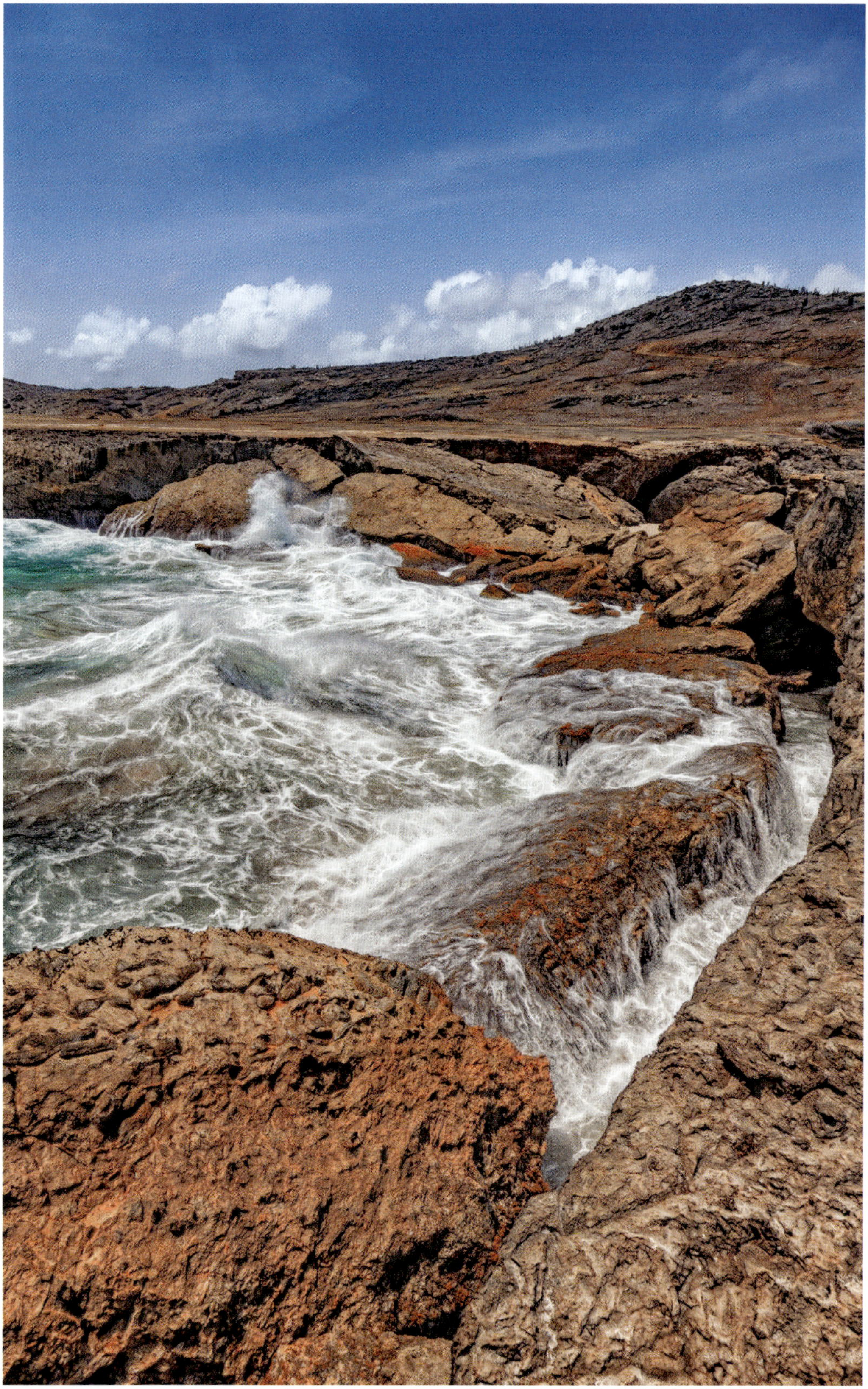

29 ARIKOK NATIONAL PARK

TO VISIT BEFORE YOU DIE BECAUSE

The natural treasures of this island park range from rocky coastline and spectacular limestone caves to cactus-studded hills.

Arikok National Park, located on Aruba in the southern Caribbean, is a natural gem that encompasses nearly 20 percent of the island's terrain. And though it's only 13 square miles, this park's landscape includes rugged terrain, dramatic cliffs, pristine beaches, and desert-like stretches. One of the most popular spots for visitors is the Fontein Cave, where Aruba's native population once left their drawings on the limestone walls. The cave is often quite warm, so afterward it's a good idea to visit the natural pools nestled amongst volcanic rock formations on the northeastern coast for a dip. Chances are good that you'll see wild donkeys and goats while in the park, though other animals, like the riotously colorful parakeets, whiptail lizards, and burrowing owls are a possibility, too, if you have patience and a sharp eye.

www.aruba.com/us/explore/arikok-national-park

THE AMERICAS — BOLIVIA

30 AMBORÓ NATIONAL PARK

TO VISIT BEFORE YOU DIE BECAUSE

With staggering numbers of native birds and mammals, including condors and pumas, this is one of the world's most biodiverse parks.

Nestled in the heart of Bolivia, between the Andean foothills and the Amazon Basin, Amboró National Park showcases a spectacular array of biodiversity. The 1,709-square-mile reserve holds more than 912 species of birds (including Andean condors, harpy eagles, white-bellied hummingbirds, and the critically endangered southern helmeted curassows) and over 177 mammals (such as pumas, ocelots, and rare spectacled bears). There are also upwards of 3,000 identified and recorded plant species. Part of the reason for the rich diversity is the varied topography: the park's altitude ranges from just under 1,000 feet to nearly 10,000 feet above sea level, with everything from lowland forests to cloud forests and cactus forests to montane shrublands. Various operators offer day trips to the park, though those really keen on immersing themselves in the sights and sounds of the jungle can opt to embark on a multi-day trek.

national-parks.org/bolivia/amboro

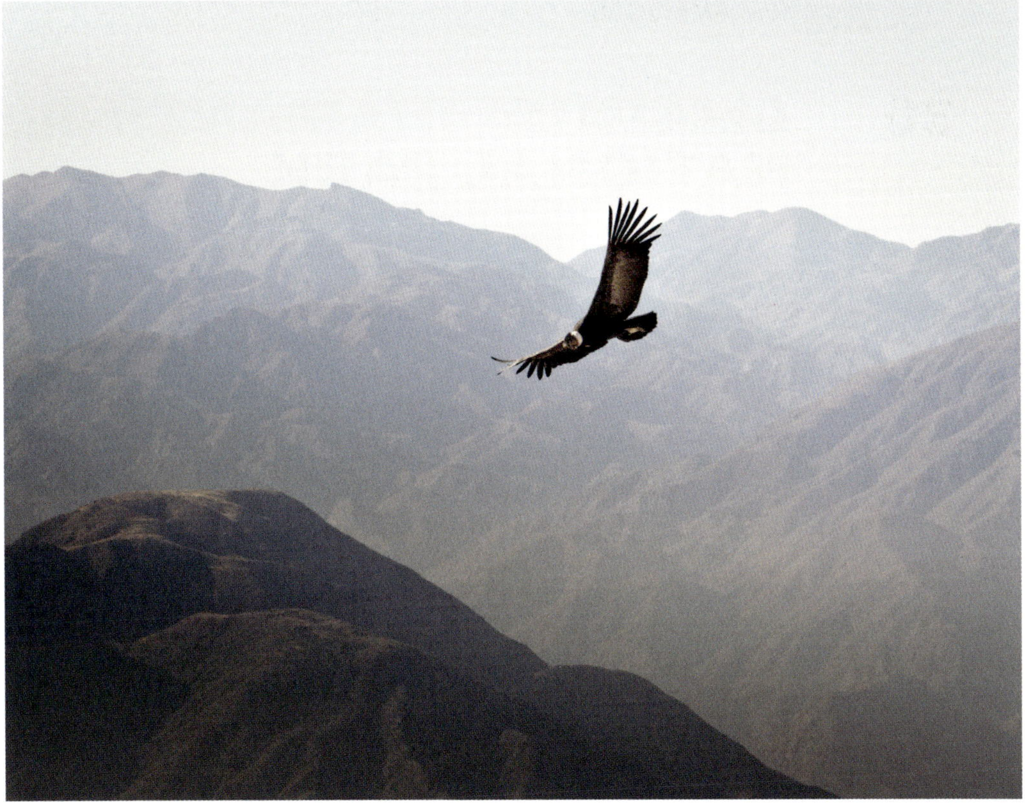

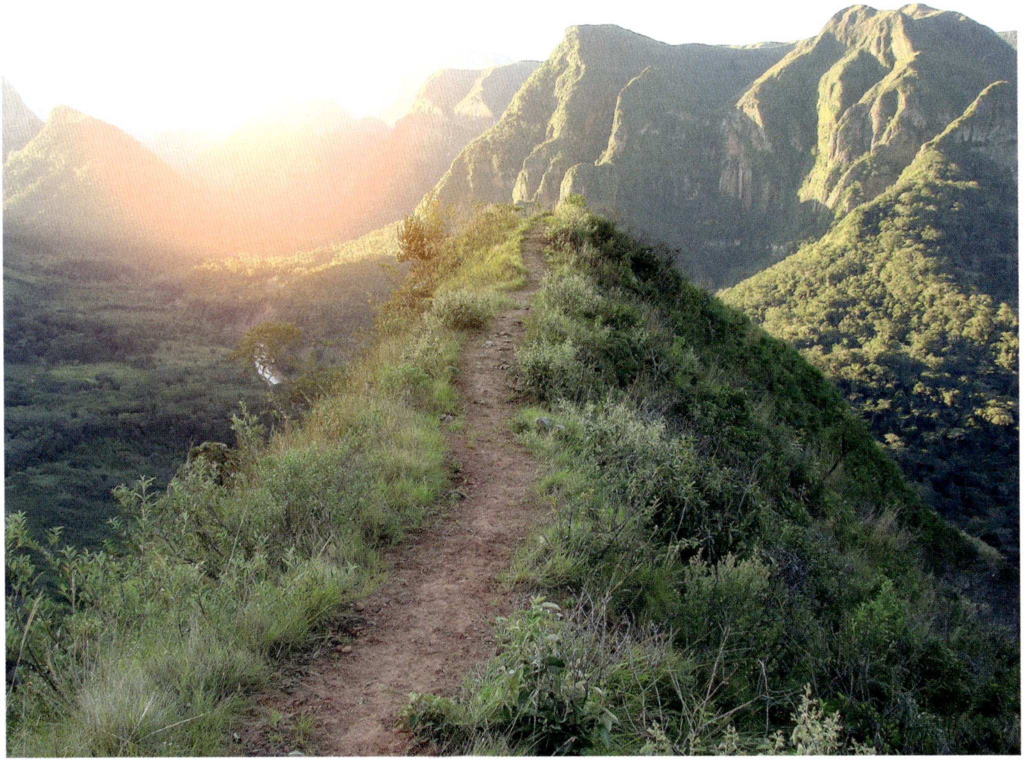

THE AMERICAS   BRAZIL

31 CHAPADA DIAMANTINA NATIONAL PARK

TO VISIT BEFORE YOU DIE BECAUSE

Embark on some epic eco-adventures in this off-the-beaten-path part of Brazil, a great destination for cave divers.

Meaning "Diamond Plateau," Chapada Diamantina was once the site of massive diamond deposits. However, diamond mining is now banned within the national park in an effort to protect some of the most beautiful mountains, valleys, caves, and waterfalls in Brazil. Chapada Diamantina has a vast network of caves, some of which are among the largest in the country, making it an exciting destination for spelunkers and cave divers. The most famous cave is Poço Azul, which features an underground pond that seems to glow blue. There are myriad activities for adventure enthusiasts beyond visiting caves, though. Chapada Diamantina offers a multitude of outdoor activities, including trekking, rock climbing, rappelling, and swimming in natural pools. The park's extensive trail system provides opportunities for both beginners and experienced hikers to explore its many wonders. One of the easiest hikes in the park is Morro do Pai Inácio, which takes you to postcard-famous rock outcroppings. Another popular option is the Cachoeira da Fumaça, which features a 1,300-foot waterfall.

www.guiachapadadiamantina.com

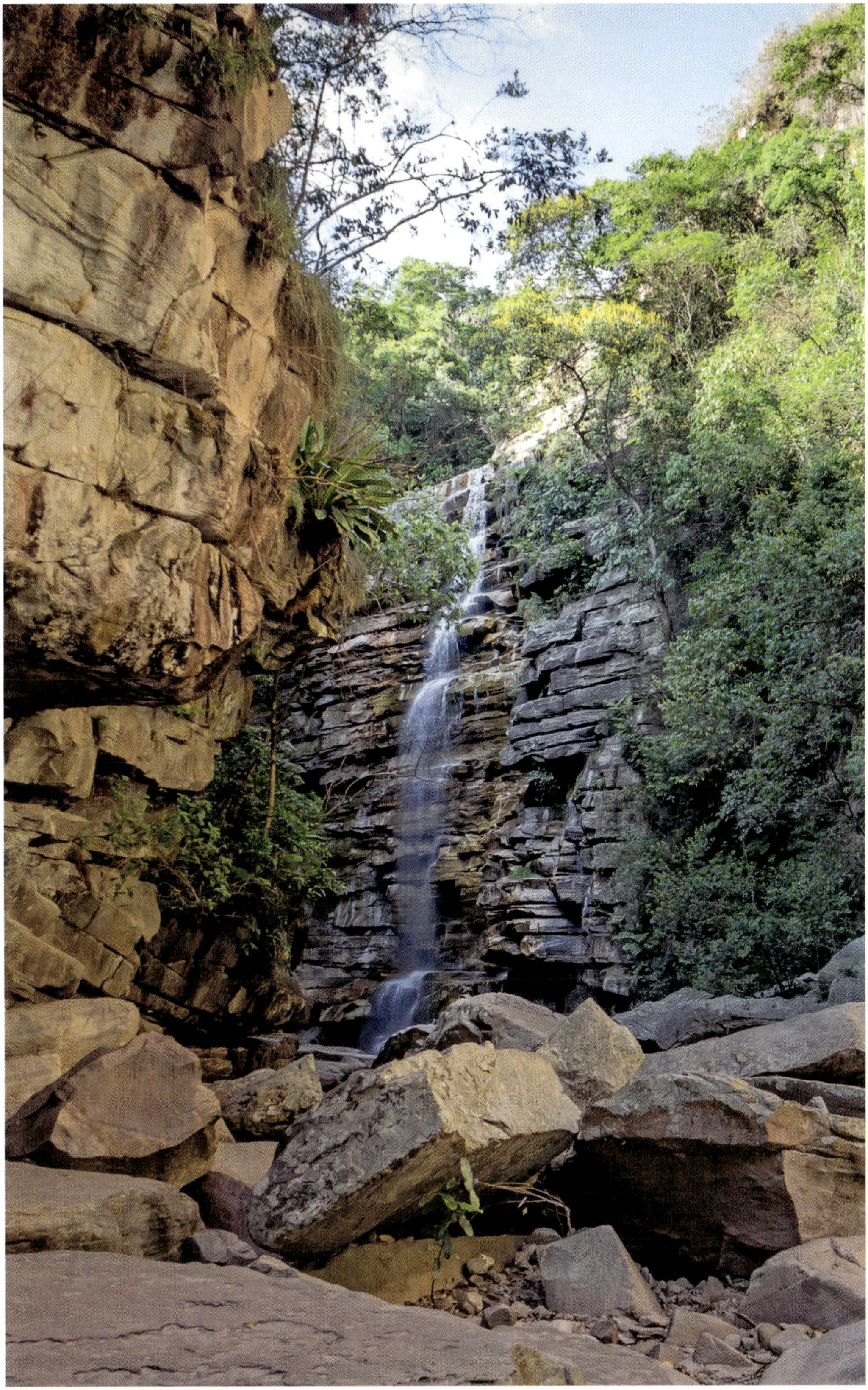

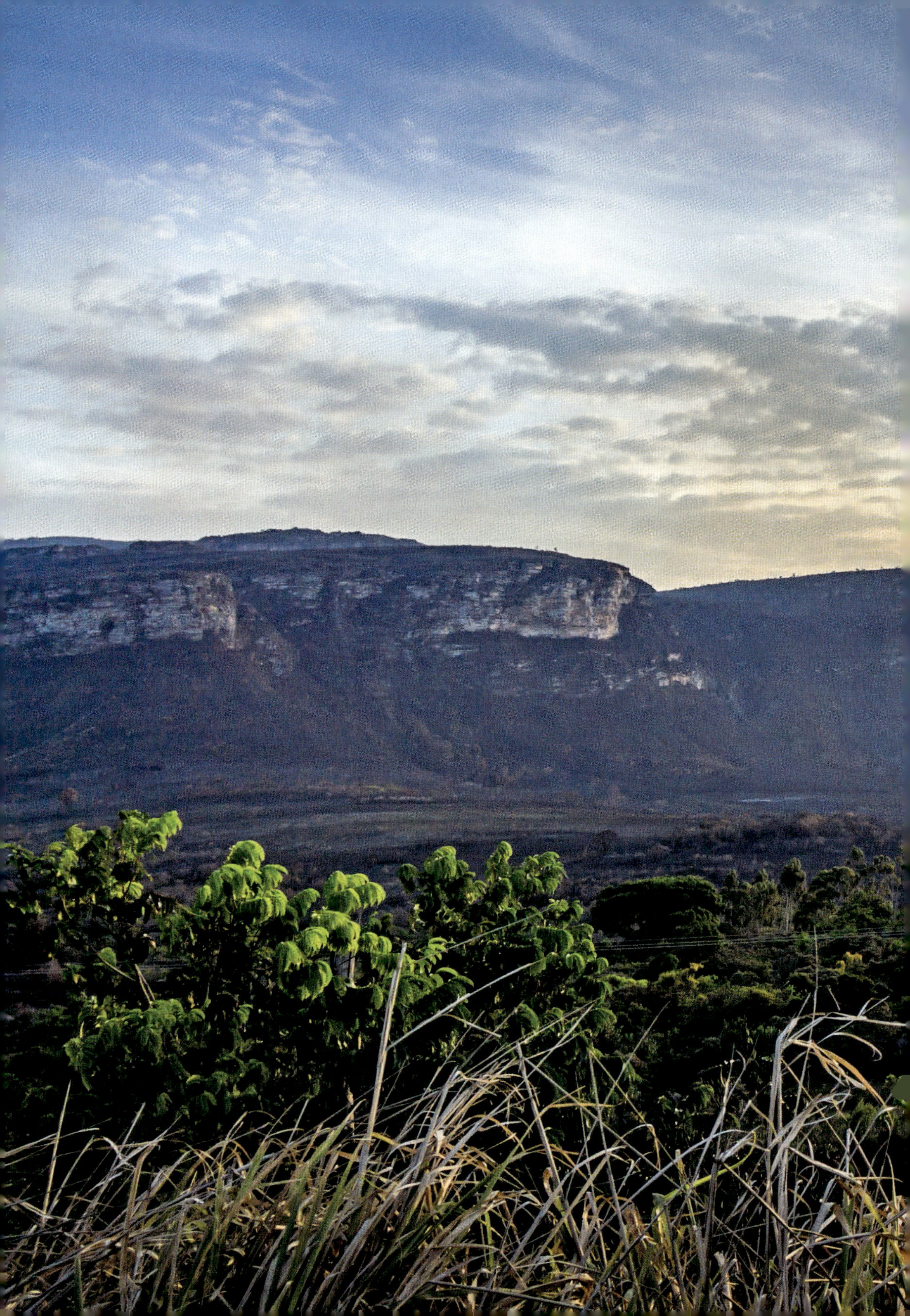

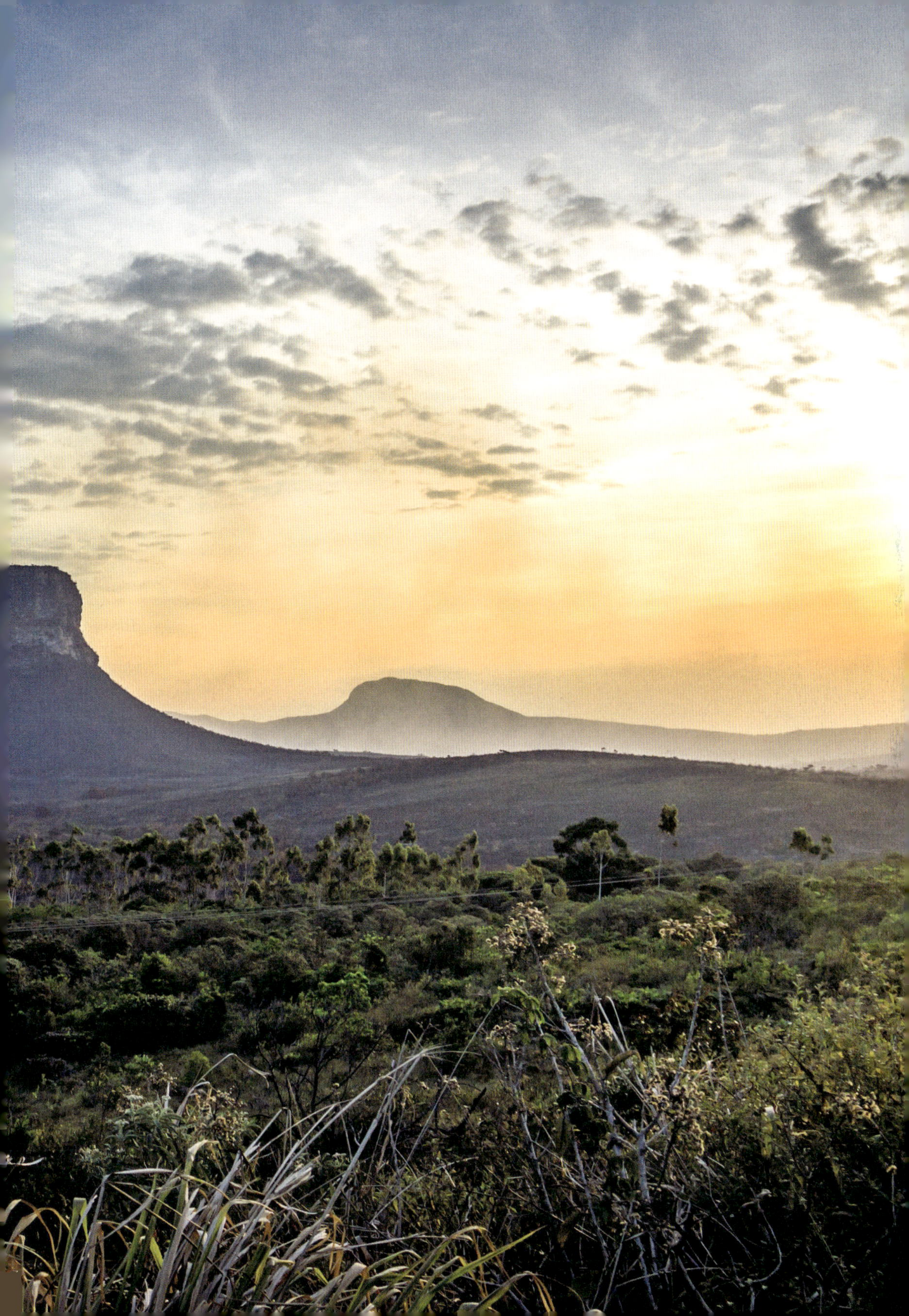

THE AMERICAS — BRAZIL

32 APARADOS DA SERRA NATIONAL PARK

TO VISIT BEFORE YOU DIE BECAUSE

When the morning mist clears away, it reveals a landscape of incredible canyons with wildlife-spotting opportunities aplenty.

Aparados da Serra National Park is a breathtaking natural reserve located in the southern region of Brazil, straddling the border between the states of Rio Grande do Sul and Santa Catarina. The park is characterized by its stunning canyons, notably the Itaimbezinho Canyon, one of the largest in Brazil, and the Fortaleza Canyon, known for its sheer rock walls that plunge into the valley below. Standing atop these imposing geological formations offers visitors unparalleled panoramic views and provides a glimpse into the region's geological history. Roughly 140 birds, 50 mammals, and 40 amphibian species have been documented in the Aparados da Serra National Park, including the mountain lion, the ocelot, the maned wolf, the green-billed toucan, the purple-bellied parrot, and the crowned eagle. Visitors to the park can seek them out on the network of hiking trails, which meander through dense forests and alongside crystal-clear streams.

parksofbrazil.mma.gov.br/serra-geral-aparados-da-serra

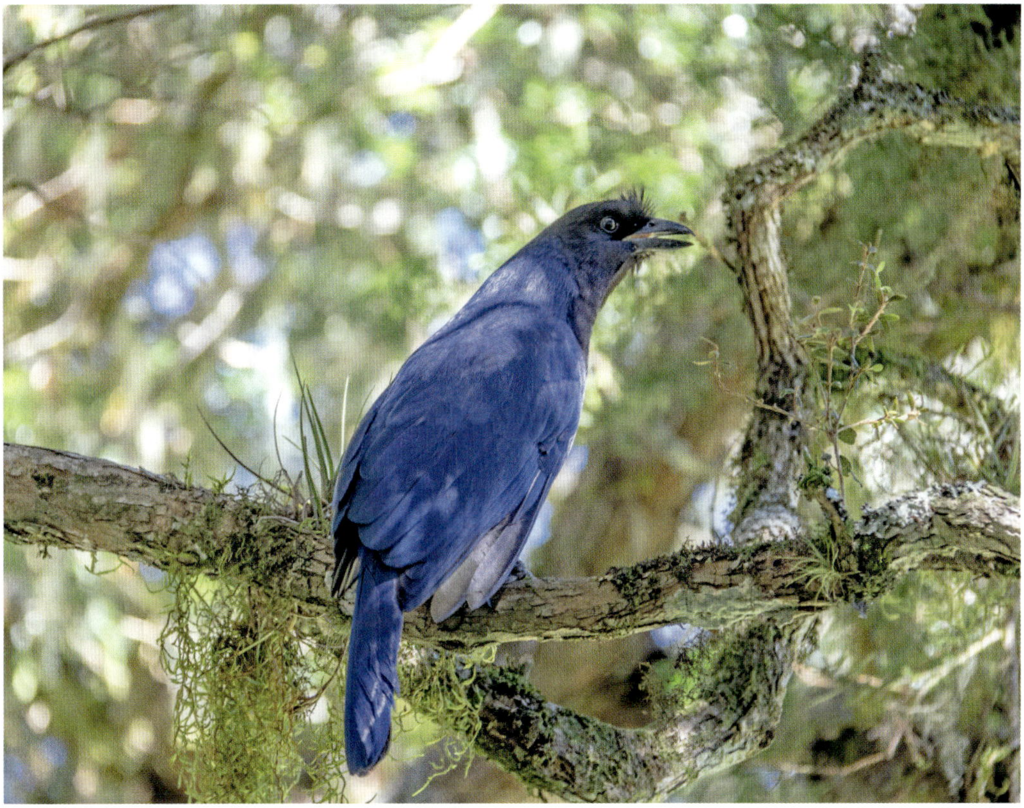

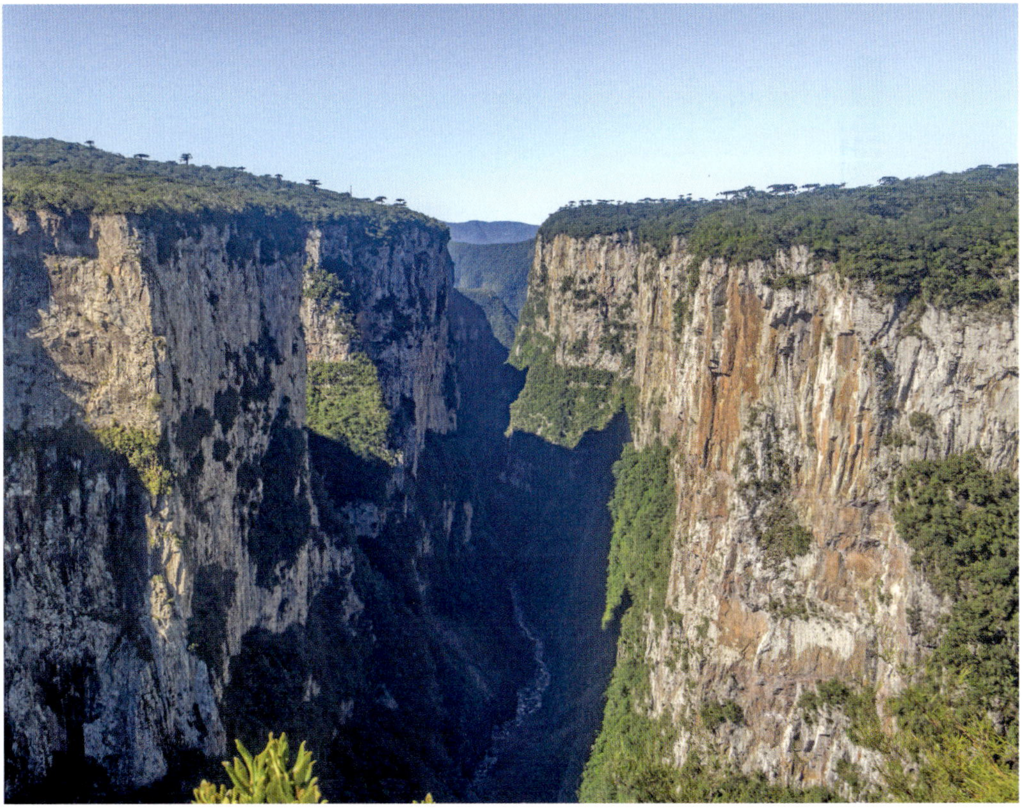

THE AMERICAS   BRAZIL

33 LENÇÓIS MARANHENSES NATIONAL PARK

TO VISIT BEFORE YOU DIE BECAUSE

This park boasts the largest field of sand dunes in South America. Interwoven with freshwater lagoons, they form a fascinating pattern.

The defining feature of Lençóis Maranhenses National Park is its surreal dune fields. Stretching as far as the eye can see, these towering dunes, some reaching heights of up to 130 feet, are sculpted by the wind into mesmerizing shapes and patterns, creating an ever-shifting landscape of remarkable beauty. Amongst the dunes are countless freshwater lagoons – at least during the rainy season, between January and June, when torrential downpours fill the low-lying areas between the dunes with crystal-clear rainwater. The resulting lagoons, with their vivid shades of blue and green, provide a stark contrast to the surrounding desert-like terrain, offering visitors the opportunity to swim and relax in this surreal setting. Interestingly, despite its appearance, the park receives too much rainfall to technically be considered a desert. Visitors to Lençóis Maranhenses National Park can explore its enchanting landscape on foot, by dune buggy, or on guided tours offered by local operators.

THE AMERICAS BRAZIL

34 FERNANDO DE NORONHA MARINE NATIONAL PARK

TO VISIT BEFORE YOU DIE BECAUSE

Located on a volcanic archipelago off Brazil's northeast coast, this park is an incredibly important habitat for turtles, tuna, and other tropical fauna.

If you love beaches and sea life, Fernando de Noronha Marine National Park in Brazil should be at the top of your bucket list. Formerly a penal colony, the Unesco World Heritage Site and national park comprises 21 islands and islets, all formed by volcanic peaks in a submerged mountain chain, with Fernando de Noronha being the largest. Beneath the surface of the azure waters, colorful coral reefs support an abundance of marine life, including tropical fish, dolphins, and even sharks. As such, it's an exciting destination for snorkeling and diving expeditions, as well as aquatic research. On land, the isles feature dramatic cliffs, mangroves, and white sand beaches, the latter of which are an important breeding site for both the hawksbill and green turtle. The islands are also home to the most significant concentration of tropical seabirds, like the red-footed booby, magnificent frigatebirds, and great skuas, in the Western Atlantic.

www.parnanoronha.com.br/en/visitante

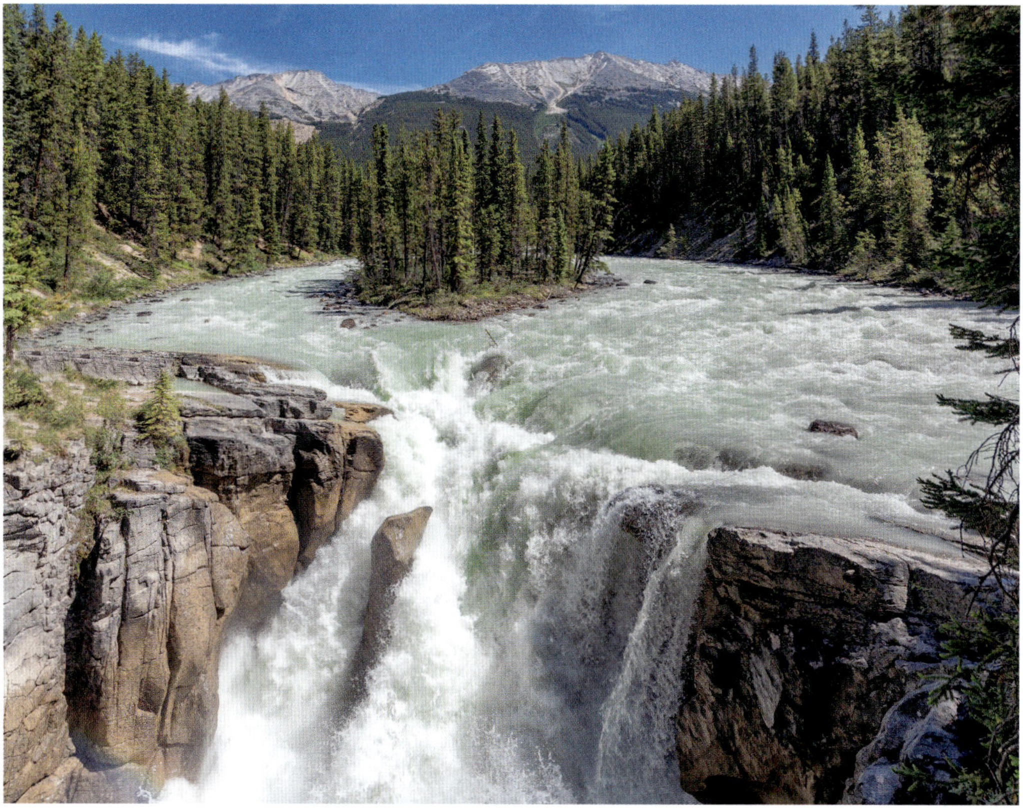

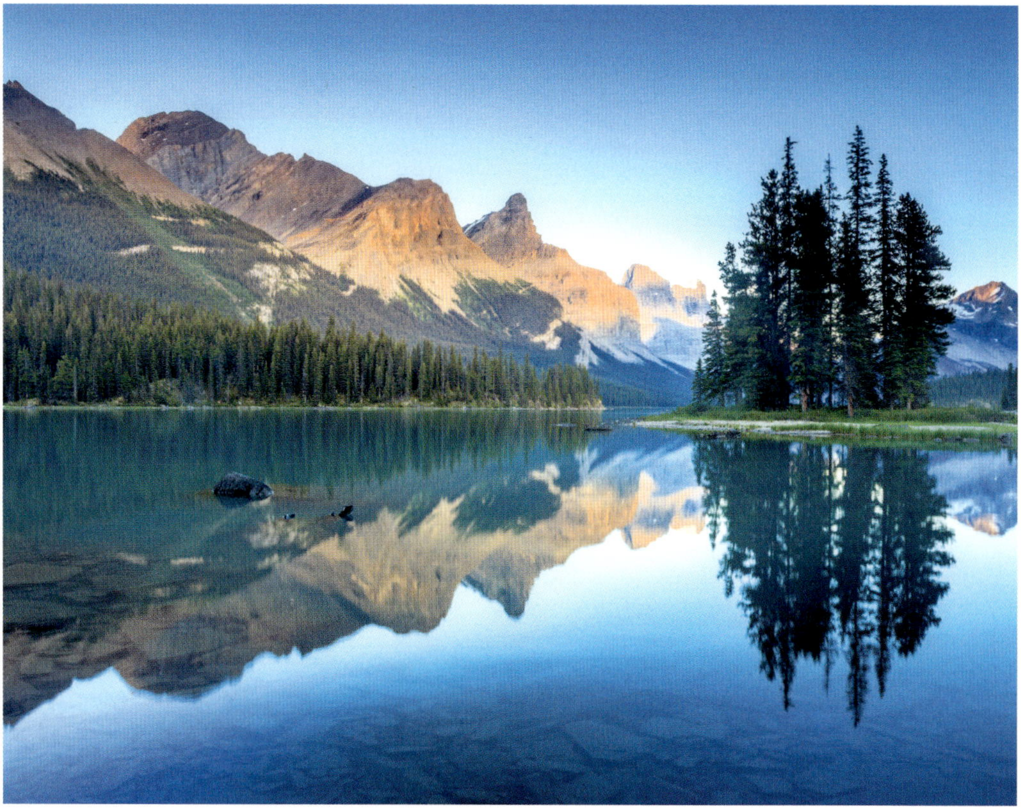

THE AMERICAS · CANADA

35 JASPER NATIONAL PARK

TO VISIT BEFORE YOU DIE BECAUSE

Walk on the wild side in the largest national park in the Canadian Rockies, and the world's second-largest dark sky preserve.

A hundred years ago, wildernesses like Jasper were commonplace in the West. Now, this protected land is amongst the last truly wild parklands in North America. Outdoor enthusiasts flock to Jasper year-round for its unparalleled recreational opportunities. In the warmer months, hiking trails crisscross the park, offering everything from leisurely strolls to challenging alpine treks. Routes like the Skyline Trail and the Bald Hills offer vistas of rugged peaks and alpine meadows carpeted with vibrant wildflowers. Adventurers can also partake in rock climbing, whitewater rafting, horseback riding, and mountain biking. As the seasons change and winter blankets the landscape in snow, the Marmot Basin ski area beckons skiers and snowboarders with its pristine powder. Meanwhile, cross-country skiing, snowshoeing, ice climbing, and ice-skating opportunities abound throughout the park. Jasper National Park is a designated dark sky preserve, making it one of the best places on Earth for stargazing. Away from the light pollution of cities, visitors can marvel at the Milky Way stretching across the night sky and, in the winter, the twirling neon ribbons of light that are the aurora borealis.

parks.canada.ca/pn-np/ab/jasper

36 FORILLON NATIONAL PARK

TO VISIT BEFORE YOU DIE BECAUSE

The first national park to open in Quebec, this area of the Gaspé Peninsula boasts incredibly diverse habitats.

Forillon National Park, nestled on the northeastern tip of the Gaspé Peninsula in Quebec, Canada, is an interesting mosaic of rugged cliffs, dense forests, picturesque coastline, sand dunes, salt marshes, natural prairies, and the eastern end of the Appalachian Mountains, all packed into just 94 square miles. Visitors can explore a network of hiking trails that wind through these ecosystems, where they might also be offered the opportunity to witness the park's denizens, including black bears, moose, lynx, red fox, snowshoe hares, mink, ermine, and migratory birds, such as razorbills, black-legged kittiwakes, great blue herons, and peregrine falcons. Alternatively, the beaches are easily accessible, so it's possible to spend the day tide-pooling, kayaking along the rocky shore, and scanning the water for signs of whales. Should you wish to stay the night, there are three camping areas (Petit-Gaspé, Des-Rosiers, and Cap-Bon-Ami) where you can pitch a tent and spend the evening contemplating the constellations.

parks.canada.ca/pn-np/qc/forillon/visit

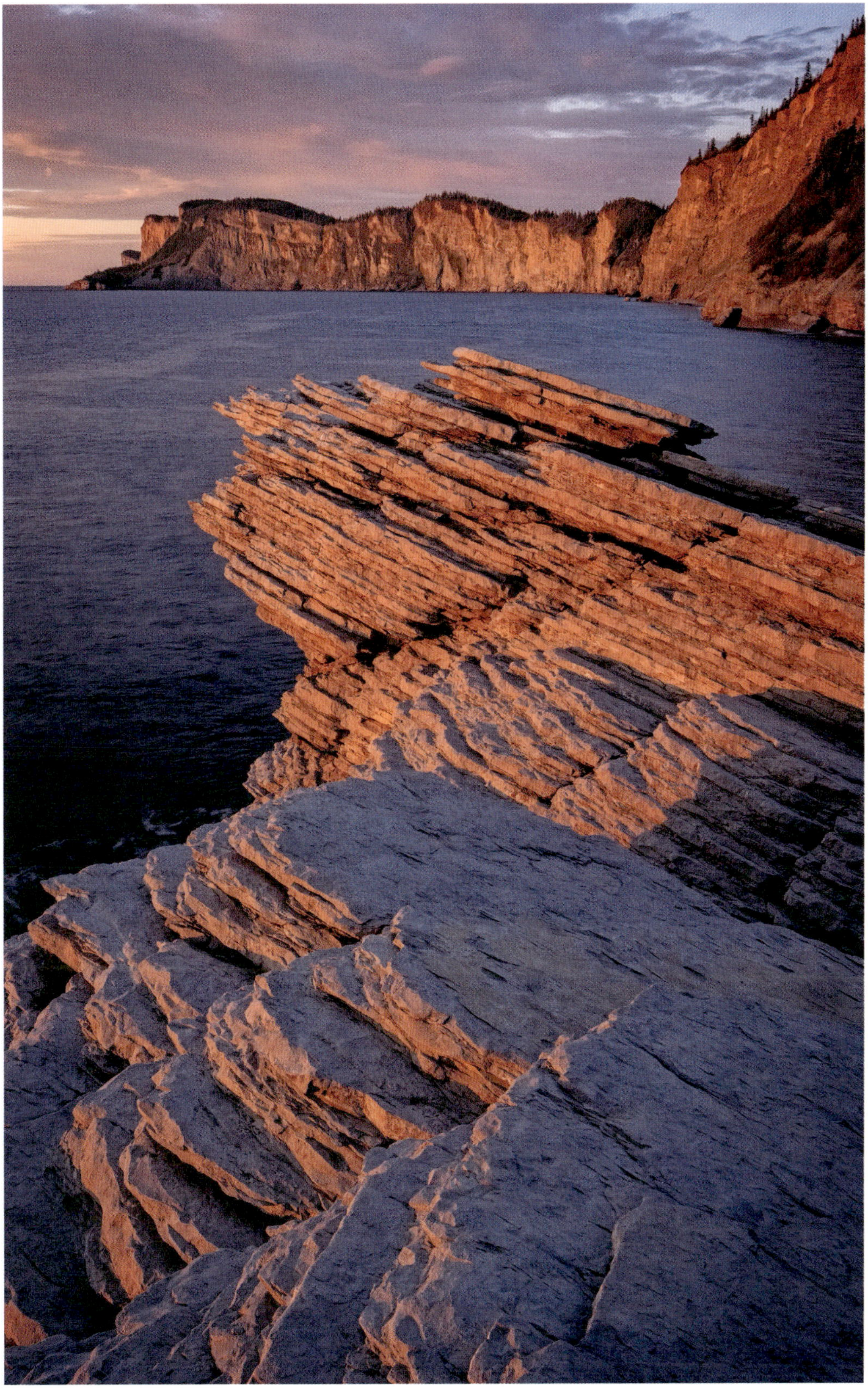

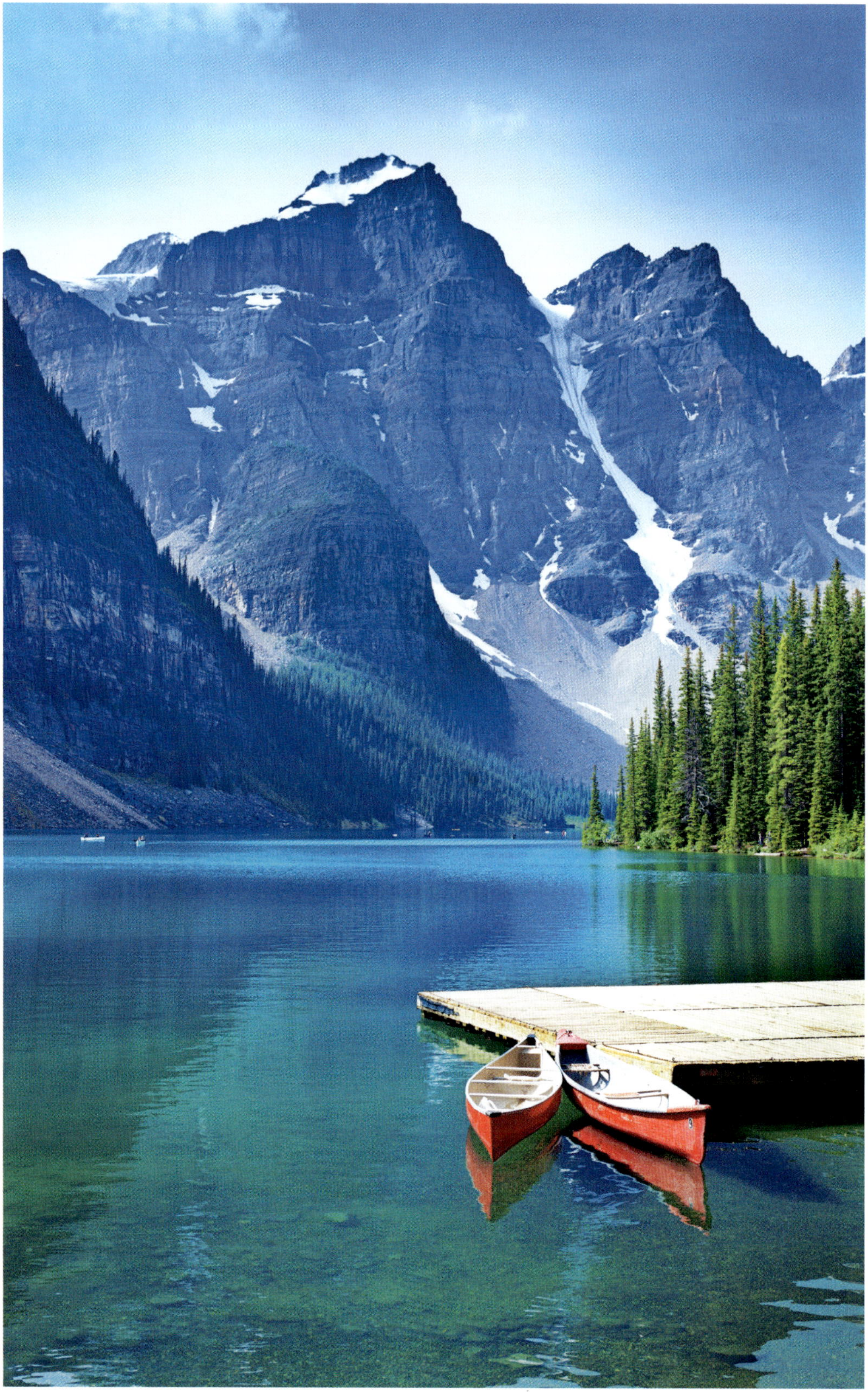

THE AMERICAS — CANADA

37 BANFF NATIONAL PARK

TO VISIT BEFORE YOU DIE BECAUSE

The flagship of Canada's park system, Banff is part of a Unesco heritage site encompassing Rocky Mountain peaks and glacial lakes.

Set amidst the breathtaking Canadian Rockies in Alberta, Banff National Park was established in 1885 but the area has been wowing visitors for much longer. With towering snow-capped peaks, glacial lakes, dense pine forests, and cascading waterfalls, the landscape of Banff seems almost surreal in its grandeur. The jewel of Banff is undoubtedly Lake Louise, its milky blue waters reflecting the surrounding peaks, including the iconic Victoria Glacier. Nearby, the equally stunning Moraine Lake captivates visitors with its vibrant turquoise hue and dramatic mountain backdrop. The park is home to a diverse array of wildlife, including grizzly bears, elk, mountain goats, and bighorn sheep. For those seeking adventure, Banff offers a plethora of outdoor activities year-round. From hiking and mountain biking in the summer to skiing and snowboarding in the winter, there's no shortage of ways to explore this natural wilderness playground.

THE AMERICAS   CHILE

38 TORRES DEL PAINE NATIONAL PARK

TO VISIT BEFORE YOU DIE BECAUSE

Head to Patagonia to admire the reflection of the mountains on Lake Pehoe, in an area sometimes known as the Eighth Wonder of the World.

Located in Chilean Patagonia, Torres del Paine National Park is the largest and most popular national park in the country. The Unesco World Biosphere Reserve stretches more than 1,500 square miles and gets its name from the three towering granite peaks, or "torres," that pierce the sky and provide a dramatic backdrop. It should come as no surprise that Torres del Paine is a paradise for hikers. The famed W Circuit trek is a bucket-list adventure, taking travelers through some of the park's most stunning scenery, including the Valle del Frances, where sharp cliffs and a collection of waterfalls create a landscape of unparalleled beauty. To experience some regions of the park you'll need a certified guide, however. But these park experts can help point out the wildlife that thrives within the park – visitors can expect to spot guanacos, foxes, Andean condors, Chilean flamingos, and Austral parakeets.

www.chile.travel/en/where-to-go/destination/torres-del-paine

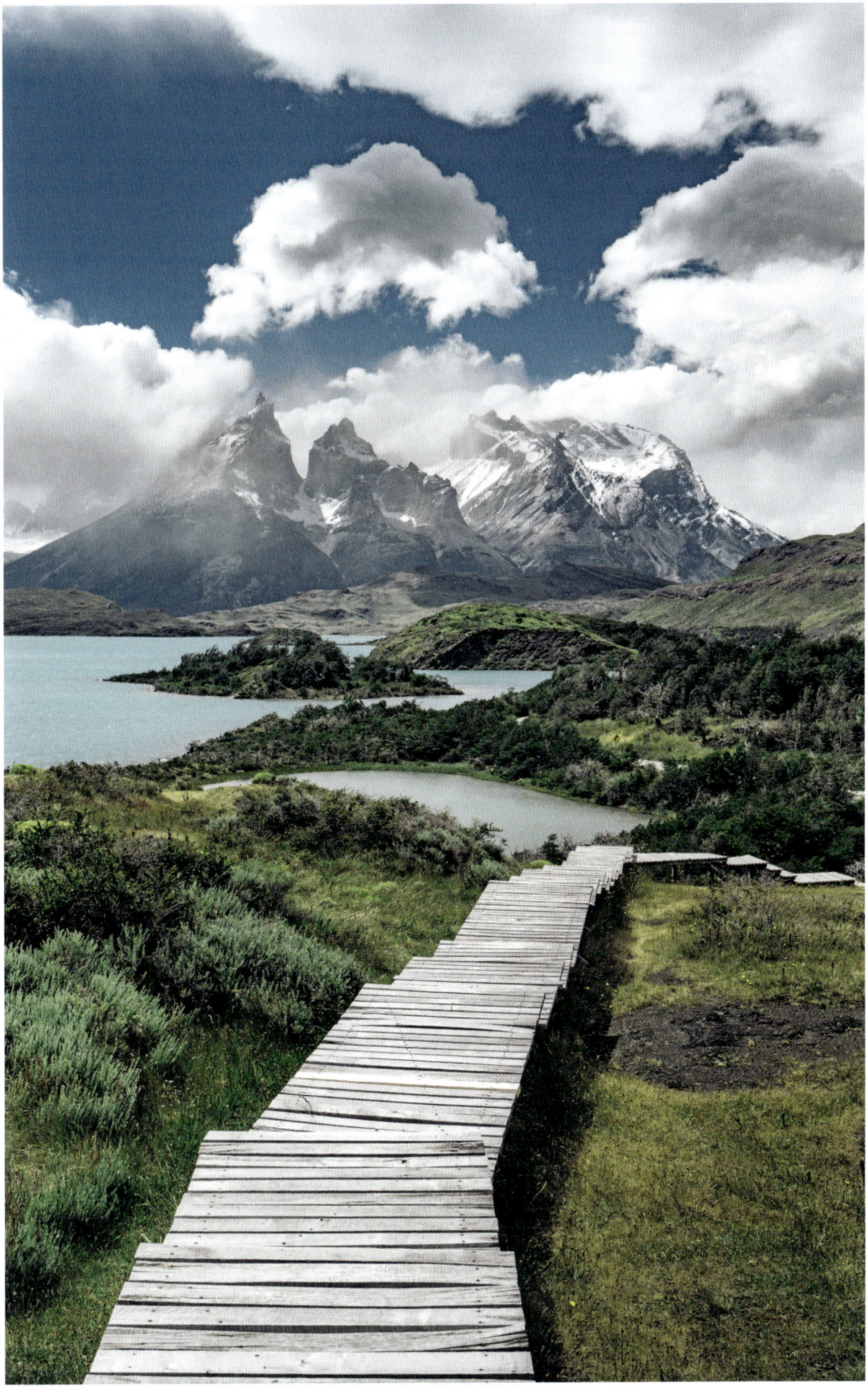

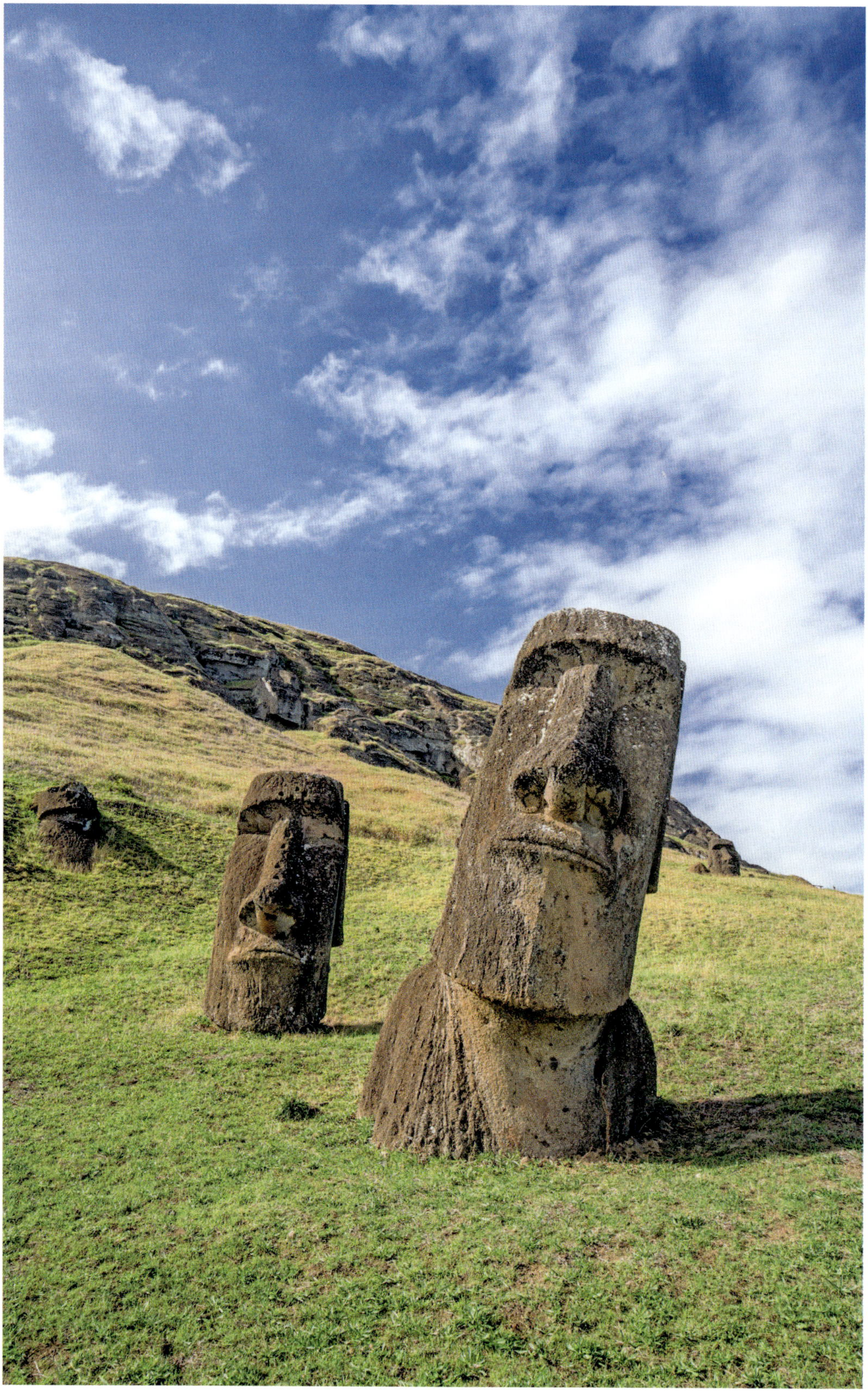

THE AMERICAS CHILE

39 RAPA NUI NATIONAL PARK

TO VISIT BEFORE YOU DIE BECAUSE

Contemplate the enduring legacy of humanity's ancient civilizations in a park now managed by the Maú Henua Indigenous Community.

Rapa Nui National Park, located on Easter Island in the southeastern Pacific Ocean, is a Unesco World Heritage Site renowned for the enormous stone statues known as moai. Carved by the island's early inhabitants, who came from Polynesia between the 13th and 16th centuries, the statues represent their ancestors and are scattered throughout the park. The massive statues were carved from volcanic tuff quarried from the island's crater rims. Then, these monolithic figures were meticulously crafted using stone tools, transported across the island's rugged terrain, and erected upon ceremonial platforms known as ahu, positioned to overlook important tribal sites and agricultural landscapes, serving as focal points for communal rituals. It's estimated that there are about 900 of these statues. Yet, the history of the moai is shrouded in enigma, with the decline of the civilization leaving behind unanswered questions about the purpose and significance of these imposing monuments. Today, visitors from around the globe come to marvel at the statues and ponder the mysteries of their creation.

rapanuinationalpark.com/en

THE AMERICAS   COSTA RICA

40 TORTUGUERO NATIONAL PARK

TO VISIT BEFORE YOU DIE BECAUSE

See three species of turtles lay their eggs on the sandy shores of this iconic park, composed of an extensive network of rivers and streams.

At Tortuguero National Park, thick jungle and meandering rivers intertwine, offering park-goers the chance to partake in a one-of-a-kind kayaking or motorboat safari. Along the banks and in the water itself, visitors can spot sloths, crowned night-herons, monkeys, caimans, macaws, turbans, kingfishers, and beyond. However, it is the turtles (including hawksbill, leatherback, and green sea turtles) that give the park its name and are the star of the show. In the right season, typically between March and October, and under the cover of darkness, visitors can watch as mother turtles lay their eggs (or see the baby turtle wriggle their way to the surf). All three species of turtles have thrived in these protected waters of the Caribbean Sea for centuries. The marine environment around the park is equally captivating, with vibrant color reefs and seascapes teeming with life. Snorkelers and divers can spot kaleidoscopic fish darting through the swaying sea fans.

www.visitcostarica.com/en/costa-rica/where-to-go/protected-areas/tortuguero-national-park

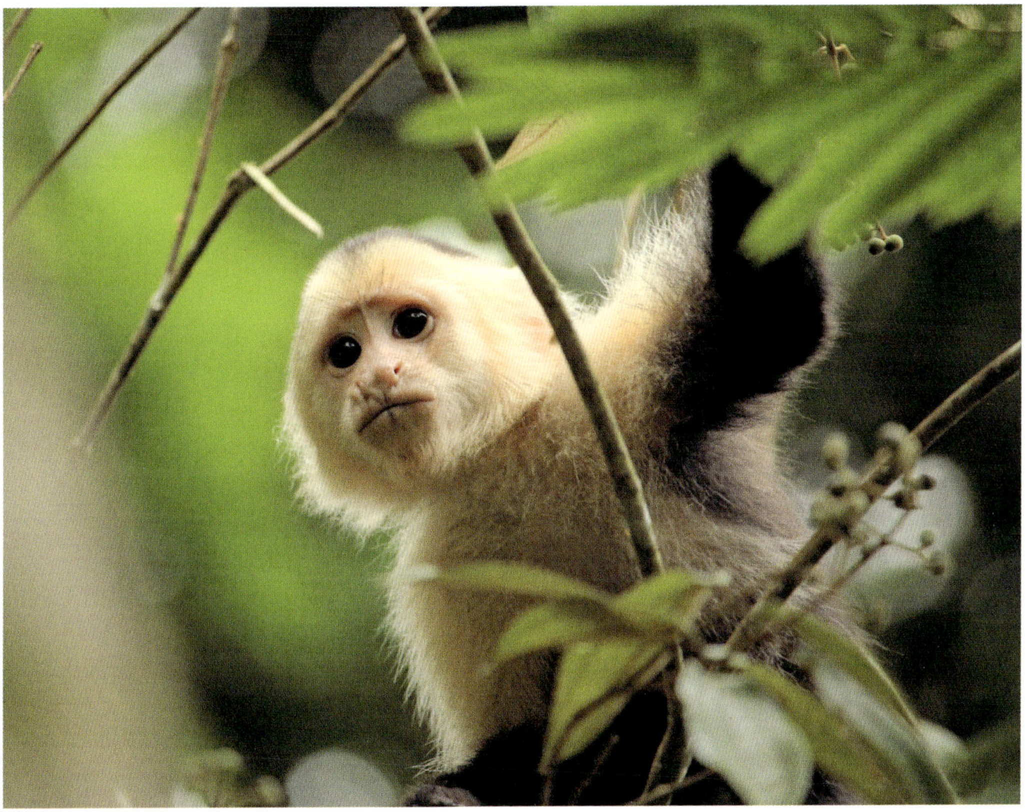

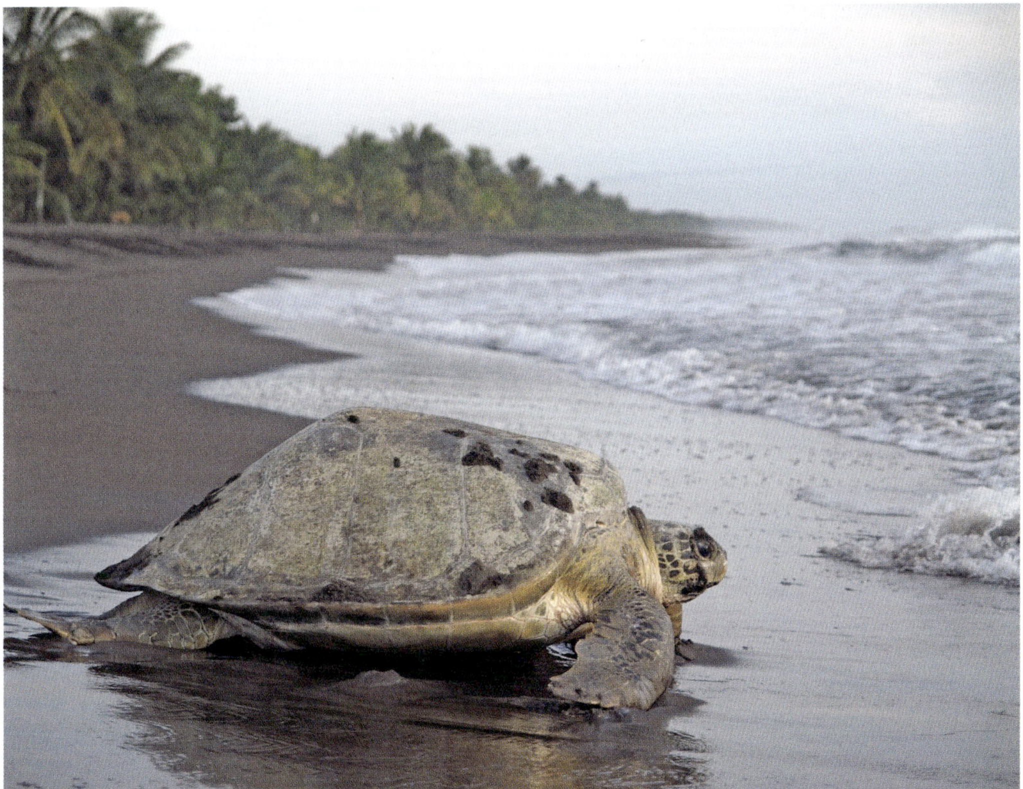

THE AMERICAS   COSTA RICA

41 CORCOVADO NATIONAL PARK

TO VISIT BEFORE YOU DIE BECAUSE

Home to 2.5 percent of the world's biodiversity, this stretch of mangroves and lagoons is one of the few remaining tropical lowland forests on the planet.

Located on the Oso Peninsula in the southwest area of Costa Rica, Corcovado National Park protects 13 major ecosystems over 164 square miles. It also hosts the largest primary forest on the American Pacific coastline. Among the biodiversity in the park, visitors can enjoy over 6,000 species of insects, 500 trees, 367 birds, 140 mammals, 117 amphibians and reptiles, and 40 freshwater fish. These include the endangered Baird's tapir and the harpy eagle. Other notable animals you may encounter are anteaters, bull sharks, crocodiles, giant sloths, otters, sea turtles, hummingbirds, frogs, hermit crabs, butterflies, monkeys, including red-backed squirrel monkeys, and several feline species including jaguar, ocelot, and pumas. To enter the park you must be accompanied by an official guide. There are four ranger stations where overnight camping is permitted. There are two major hiking tracks within the park, one interior and one coastal. The coastal trail involves several river crossings that are inhabited by crocodiles and caiman, so it is recommended only for experienced hikers.

www.visitcostarica.com/en/costa-rica/where-to-go/protected-areas/corcovado-national-park

THE AMERICAS   ECUADOR

42 GALÁPAGOS NATIONAL PARK

TO VISIT
BEFORE YOU DIE
BECAUSE

It boasts large numbers of endemic, critically endangered species that were the inspiration for Charles Darwin's Theory of Evolution.

A Unesco World Heritage Site, Galápagos National Park makes up 97 percent of the Galápagos Islands. Established in 1959, the park protects 3,087 square miles of land and over 50,000 square miles of surrounding waters, known as the Galápagos Marine Reserve. The volcanic archipelago consists of 18 main islands, though there are only four inhabited areas where visitors can roam freely – they need to be with a certified naturalist guide to visit the more far-flung regions. Typically, guests base themselves in one of the two main towns and take day trips around the islands or participate in multi-day expedition-style cruises. Notable species endemic to the islands include the Galápagos hawk, Galápagos sea lion, Galápagos land iguana, and the Galápagos green turtle. Other local species include the blue-footed booby, swordfish, manta ray, marine iguana, waved albatross, flightless cormorant, and hammerhead sharks. The animals are famously curious about humans because they had no predators for millions of years. There are also the famous finches. In the 1830s, Charles Darwin noticed the finches on each island had slightly different beaks, perfectly adapted to the food sources available at each location. These differences were the impetus behind Darwin's 1859 book On *the Origin of Species*, which is considered the foundation of modern evolutionary biology.

areasprotegidas.ambiente.gob.ec/en/areas-protegidas/galapagos-national-park

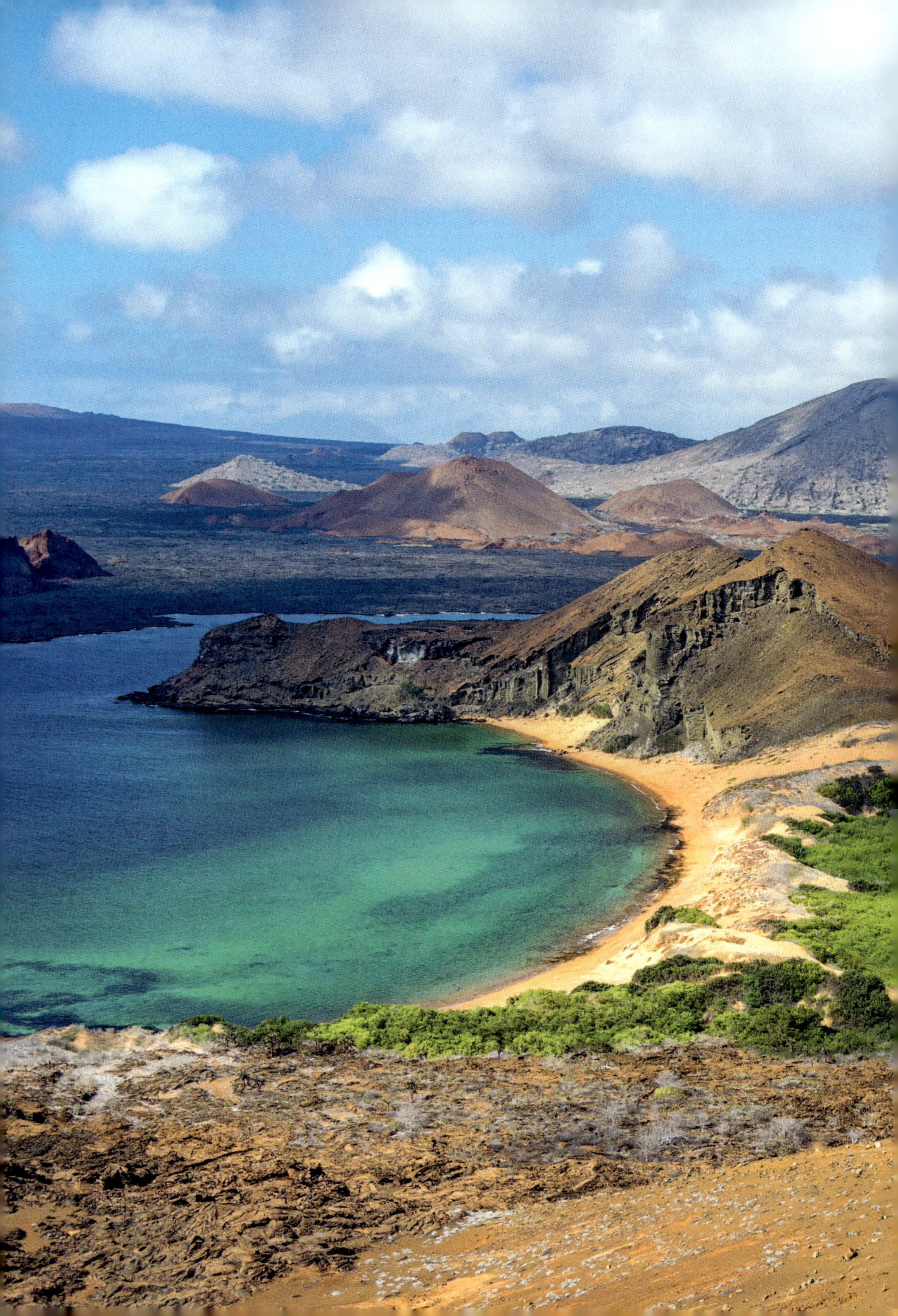

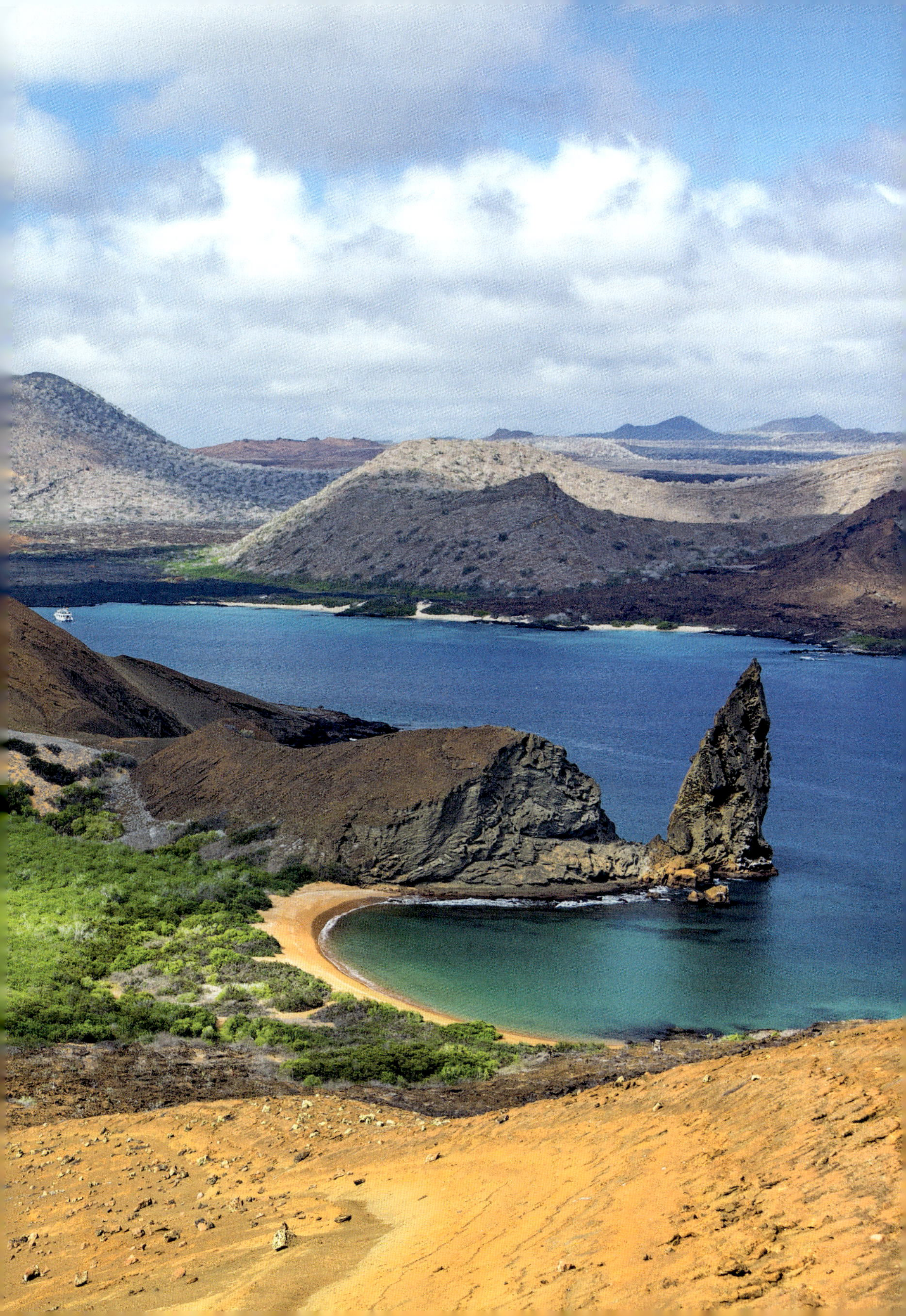

THE AMERICAS   ECUADOR

43   COTOPAXI NATIONAL PARK

TO VISIT BEFORE YOU DIE BECAUSE

This popular natural reserve is home to the snow-capped Cotopaxi, one of the world's highest active volcanoes.

At just under 130 square miles, Cotopaxi National Park offers a wide variety of activities for visitors, including hiking, camping, horseback riding, mountain biking, and mountaineering. Hiking along the Rio Pita will provide you views of numerous waterfalls, as well as more than 500-year-old homes built by native Ecuadorians before the Spanish invasion of the 16th century. Wildlife you're likely to encounter include deer, fox, llamas, wolves, marsupial mice, pumas, myriad birds, and over 200 plant species. If you're lucky you might catch a glimpse of the Andean spectacled bear, a vulnerable animal that is the only extant species of bear native to South America. With its peak elevation of 19,348 feet, the park offers various vegetation zones at different altitudes. The volcano's peak is snow-capped year round, preventing vegetation from growing at the top. The snow gives way to a humid mountain forest as you decrease in elevation. Below that resides rainy, sub-Andean plains rife with grasses and mosses.

THE AMERICAS   GUATEMALA

44   TIKAL NATIONAL PARK

TO VISIT
BEFORE YOU DIE
BECAUSE

Get a taste for what it was like to live during the Mayan Empire at the ancient city of Tikal, surrounded by tropical forests, savannahs, and wetlands.

Located in the heart of the rainforest of northern Guatemala, this Unesco World Heritage Site is one of the most iconic archaeological areas of the ancient Maya civilization, with towering pyramids, temples, and intricate stone carving that bear witness to the ancient culture. Dating back to around 800 BC, Tikal flourished as a powerful city-state and cultural center, reaching its peak during the Classic period (200–900 CE). At its height, it was one of the largest and most influential cities in the Mayan world, with an estimated 100,000 inhabitants. Today the ruins provide insight into the religion, politics, and daily life of the Mayans. Beyond the former city, the park stretches across 57,600 hectares and is also home to a vast array of wildlife. Be sure to keep an eye out for the howler monkeys as they swing through the treetops and watch for colorful toucans soaring through the canopies.

whc.unesco.org/en/list/64

THE AMERICAS   GUYANA

45 KAIETEUR NATIONAL PARK

TO VISIT BEFORE YOU DIE BECAUSE

The only national park in the country, it boasts the magnificent Kaieteur Falls, the largest single-drop waterfall anywhere on Earth.

Deep within the heart of Guyana's Amazon rainforest is Kaieteur National Park, the first and only national park in the South American country. The centerpiece of the park is undoubtedly Kaieteur Falls, a natural wonder that commands attention with its sheer size and power. The falls plunge over a sandstone cliff, forming a single, uninterrupted cascade – when it crashes into the gorge below, it creates a thunderous roar that can be heard from miles away. At 741 feet, it's about four and a half times the height of Niagara Falls, and roughly twice the size of Victoria Falls. In addition to Kaieteur Falls, the park boasts a diverse range of ecosystems, including rivers, rainforests, and savannahs. These habitats support an incredible array of plant and animal life, including endemic species such as the golden frog and the largest butterfly in South America, as well as iconic Amazonian creatures like the giant otter, the harpy eagle, and the riotously colorful Guianan cock-of-the-rock.

www.guyanapnc.org/kaieteur-national-park

THE AMERICAS   MEXICO

46 TULUM NATIONAL PARK

TO VISIT BEFORE YOU DIE BECAUSE

See some of the best Mayan ruins in Mexico before exploring Mexico's Carribbean coastline and Yucatán Peninsula.

The centerpiece of Tulum National Park is the archaeological site of Tulum, which dates back to the 13th century and was once an important trading port for the Mayan people. The city's well-preserved ruins, which include temples, ceremonial structures, stone pathways, and walled fortresses, sit on 40-foot cliffs overlooking the Caribbean Sea. It was one of the last cities built and inhabited by the Mayans and was abandoned sometime around the 16th century. Because the site is so well-preserved, it's one of the most popular archaeological sites in Mexico – only Teotihuacán and Chichen Itza see more annual visitors. Beyond its man-made treasures, Tulum National Park is also blessed with breathtaking natural beauty, including pristine beaches, coastal mangroves, and riotously colorful coral reefs. The clear waters just off the shore are excellent for swimming and snorkeling and the mangroves are ripe for spotting tropical birds and iguanas.

descubreanp.conanp.gob.mx/swb/conanp/ANP?suri=169

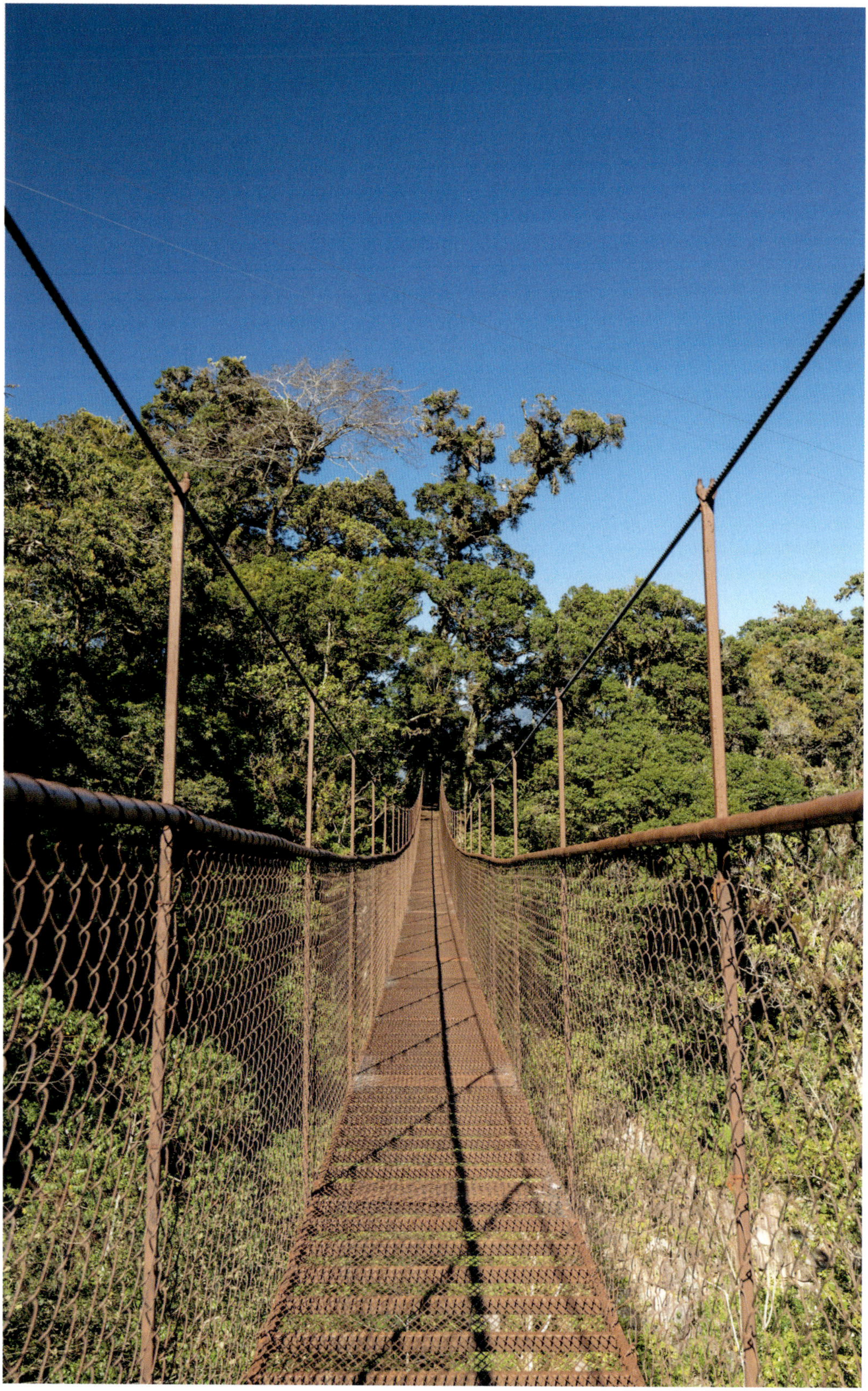

THE AMERICAS · PANAMA

47 VOLCAN BARU NATIONAL PARK

TO VISIT BEFORE YOU DIE BECAUSE

On a clear day, you can take in one-of-a-kind views of two oceans at the same time, from the highest peak in Central America.

Not only is this park's namesake peak the tallest in Panama at 11,398 feet above sea level, it's also an active stratovolcano. And given its location on the Panama isthmus, if you were standing at the summit, you might be able to see both the Pacific Ocean and the Atlantic Ocean on a clear day. There are two trails to the summit: the first, starting at the town of Volcan, is steep and strenuous, passing through thick cloud forests and crossing rocky volcanic terrain with the round trip taking about eight hours. The second, via Boquete, is accessible by 4x4 or a long but easier trail. There are myriad other hikes within the park, too. Some of the more popular ones take visitors to the seven craters of Baru Volcano – the trails vary in accessibility and degree of challenge. But along the way, visitors can keep an eye out for the more than 250 avian species that have been identified in the park, including the black-and-white hawk eagle and the wrenthrush.

www.tourismpanama.com/nature-and-parks/national-parks/volcan-baru-national-park

THE AMERICAS   PERU

48 MANU NATIONAL PARK

TO VISIT BEFORE YOU DIE BECAUSE

Experience the wild Peruvian Amazon in this huge park home to caimans, jaguars, armadillos, and over a thousand species of butterflies.

Spanning over 1.5 million hectares, this Unesco World Heritage Site and national park encompasses a unique range of ecosystems, from lowland rainforests to Andean cloud forests, making it one of the most biologically diverse places on the planet. Manu National Park is renowned for its incredible wildlife, harboring thousands of plant species and a staggering diversity of animal life. It is home to iconic species such as the jaguar, giant otter, harpy eagle, and Andean spectacled bear, as well as numerous species of monkeys, birds, reptiles, and amphibians. The park's remote and pristine nature offers refuge to many endangered and endemic species, making it a crucial stronghold for conservation efforts in the Amazon. Visitors to Manu National Park have the opportunity to explore its raw wilderness through a variety of activities, including guided hikes, boat tours along the Madre de Dios and Manu rivers, and wildlife watching excursions.

www.peru.travel/en/attractions/manu-national-park

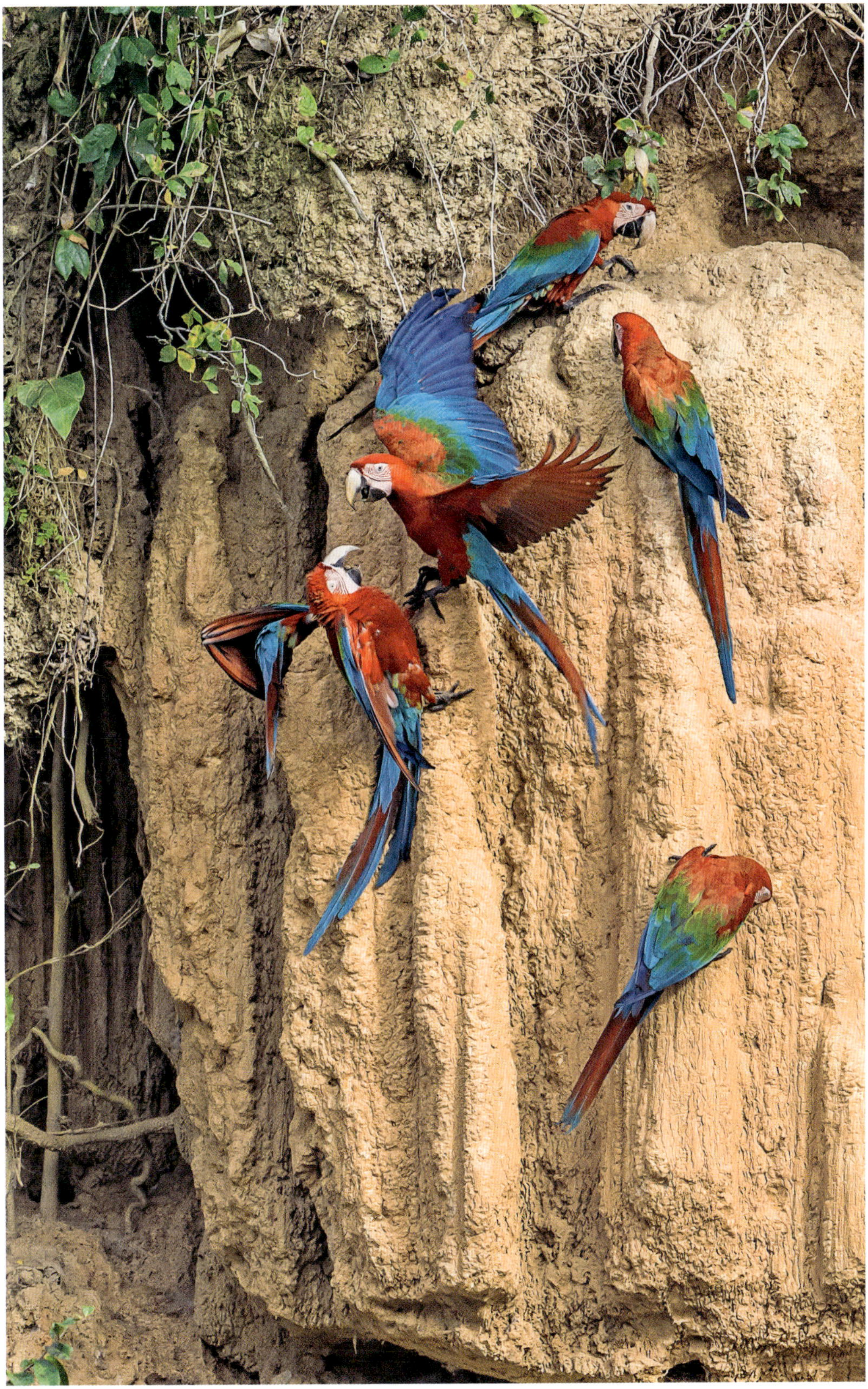

THE AMERICAS UNITED STATES (WASHINGTON)

49 MOUNT RAINIER NATIONAL PARK

TO VISIT BEFORE YOU DIE BECAUSE

Subalpine wildflower meadows and ancient forests ring the icy volcano of Mount Rainier, one of the most active volcanoes in Washington state.

Standing 14,410 feet tall, Mount Rainier is a sight to behold. The highest point in Washington state, the stratovolcano is cloaked in glaciers and surrounded by lush forests of Douglas fir, western red cedar, and western hemlock. Though it hasn't erupted since 1450, Mount Rainier is considered an active volcano. It's not a question of if it will erupt but when. Still, it serves as a beacon for outdoor enthusiasts and nature lovers alike. The National Park Service estimates that roughly 10,000 people attempt to summit the mountain each year (only half of which succeed). No matter your skill level, though, there's a hike for you within the park. Some of the most popular options include the Wonderland Trail, a 93-mile circumnavigation of the mountain, and the Skyline Trail, which offers views of cascading waterfalls, electric-blue glaciers, and colorful subalpine meadows dotted with lupines and scarlet paintbrush flowers.

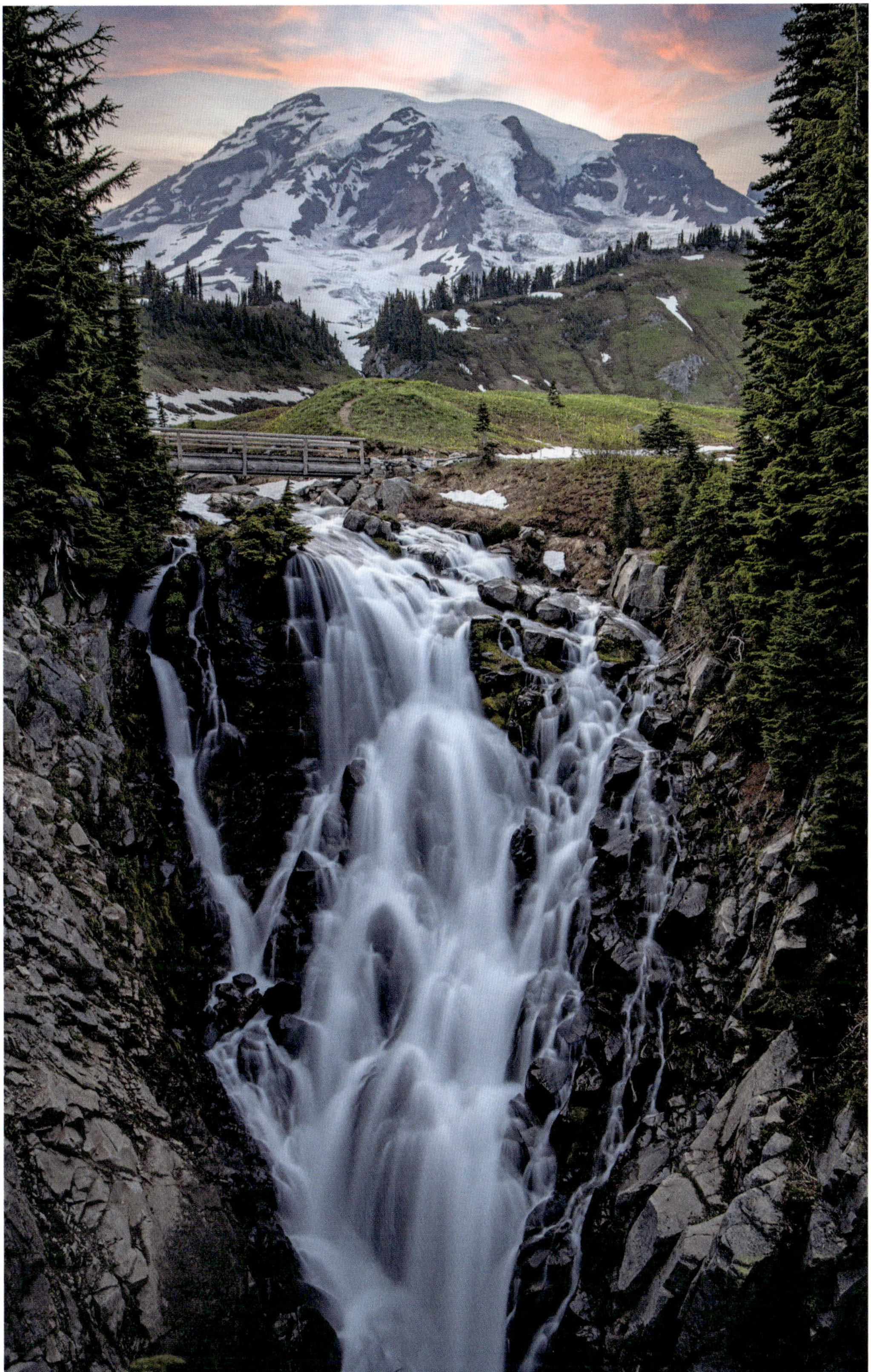

THE AMERICAS UNITED STATES (WEST VIRGINIA)

50 NEW RIVER GORGE NATIONAL PARK

TO VISIT BEFORE YOU DIE BECAUSE

A rugged, whitewater river flowing northward through deep canyons, the New River is among the oldest rivers on the continent.

Fifty-three miles of the New River run through New River Gorge National Park in West Virginia. While it doesn't have the same name recognition as other great waterways (and though its name may suggest otherwise), the fact remains that it's one of the oldest channels in the world. Over millions of years, the raging waters have dug the canyon, which now features towering sandstone cliffs, some rising over 1,000 feet. The steep sides frame the river below, creating a dramatic backdrop. The New River itself provides endless opportunities for adventure, with world-class whitewater rafting and kayaking experiences available for thrill-seekers. The river's rapids range from gentle Class I floats to exhilarating Class V challenges, ensuring there's something for everyone. Visitors looking for something a little drier, but no less thrilling can walk along the 24-inch-wide catwalk of the New River Gorge Bridge. It's one of the highest bridges in North America and spans the vast chasm above the river.

www.nps.gov/neri/index.htm

THE AMERICAS   UNITED STATES (MINNESOTA)

51 VOYAGEURS NATIONAL PARK

TO VISIT BEFORE YOU DIE BECAUSE

Paddle amongst blue heron and loons in traditional canoes in this aquatic wonderland set between southern boreal and northern hardwood forests.

America's Midwest doesn't have many national parks, but as Voyageurs shows, it should. Here, 4 major lakes and 26 smaller inland lakes make up roughly a third of the park's more than 200,000 acres, located along the northern border of Minnesota. All of its campsites (of which there are more than 200) are only accessible by water, making it an exciting place for those keen on exploring the park by motorized boat, canoe, kayak, or stand-up paddleboard. In the summer, park rangers also lead cruises on the lakes. Visitors can also sign up to participate in the North Canoe Voyage, wherein they help paddle a 26-foot canoe similar to the ones local Indigenous populations once used to travel to Canada to trade goods. Surrounding those lakes are expansive forests of pine, fir, and birch trees, which provide habitats for wolves, bald eagles, black bears, moose, and deer. The park's dark skies also make it an ideal destination for stargazing, offering unparalleled views of the cosmos above – in the winter, it may even be possible to see the northern lights.

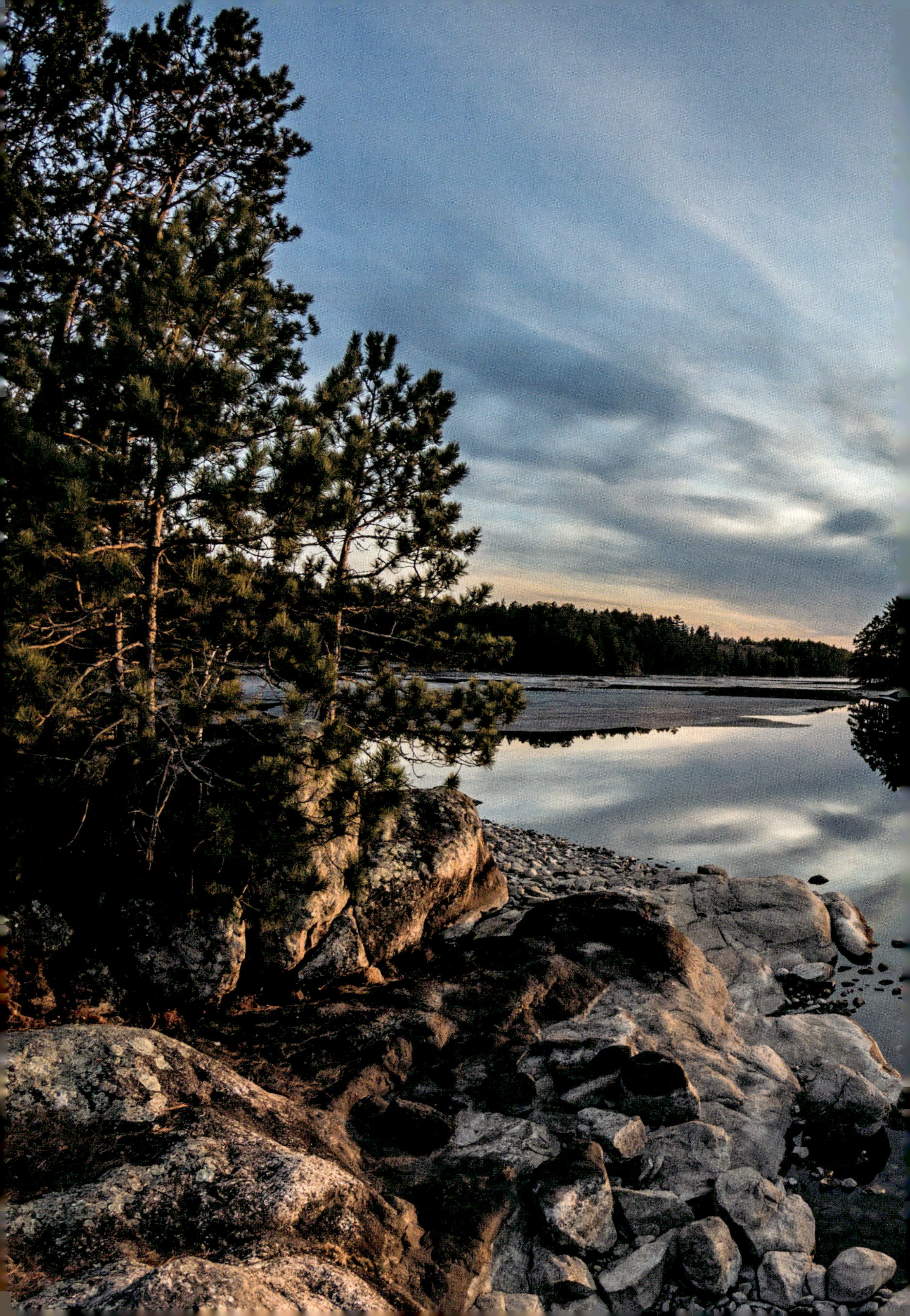

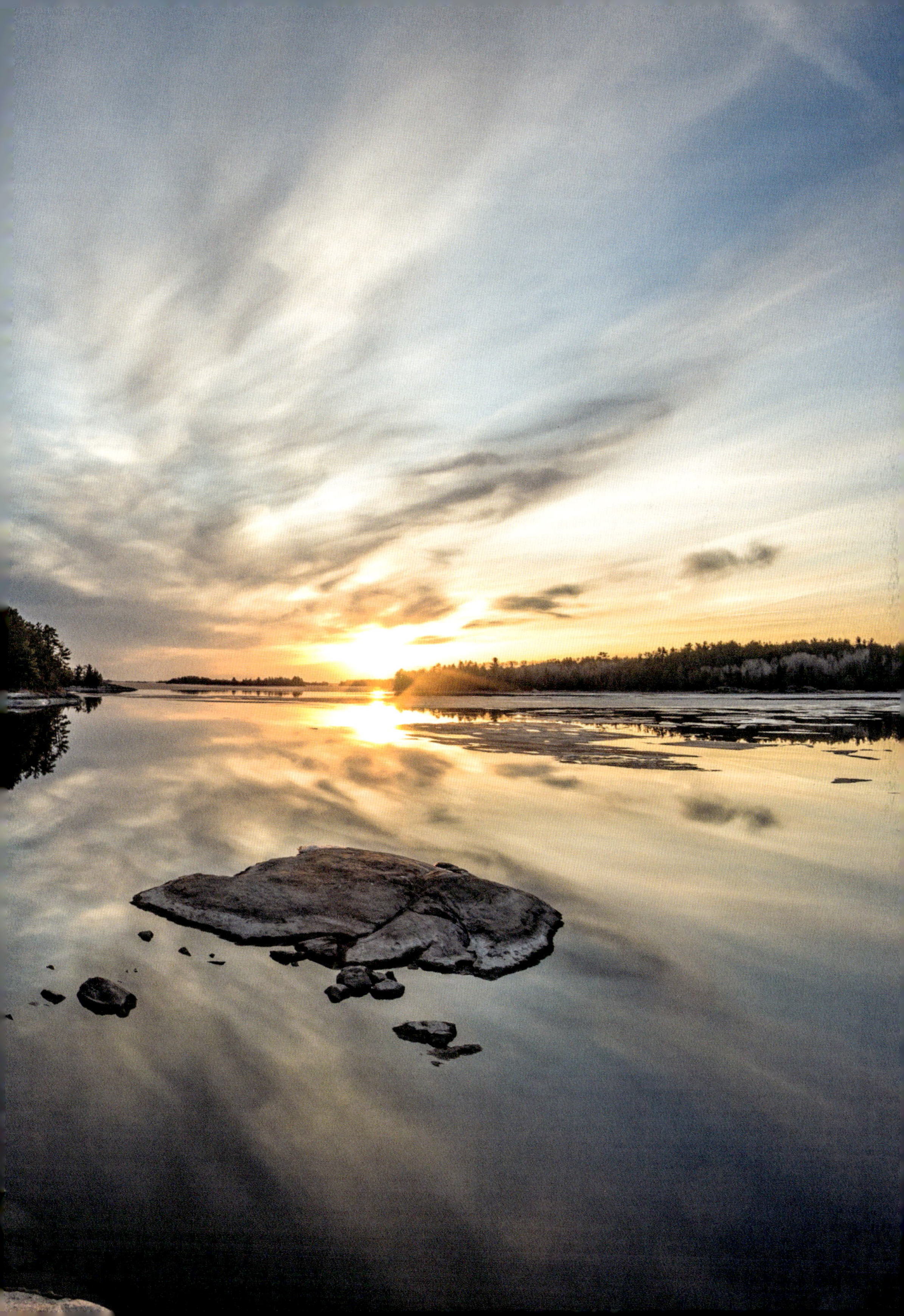

THE AMERICAS UNITED STATES (NEW MEXICO)

52 WHITE SANDS NATIONAL PARK

TO VISIT
BEFORE YOU DIE
BECAUSE

The world's largest gypsum sand dunes are found here, a glistening vision rising from the heart of the Tularosa Basin.

Spread across nearly 275 square miles of the Chihuahuan Desert, this park is renowned for its vast expanse of dazzling sand dunes. The sand is incredibly fine and pure, and because it's composed of gypsum crystals (a type of calcium sulfate), it has a brilliant white hue that sparkles under the sun. One of the park's most popular activities is the scenic eight-mile Dunes Drive, followed by sledding down the soft slopes of the dunes in plastic saucers and hiking along the handful of short trails, including the one-mile Dune Life Nature Trail and the 0.4-mile Interdune Boardwalk. In addition to its natural beauty, White Sands National Park is also rich in history and biodiversity. Ancient fossilized tracks reveal the presence of creatures that once roamed this land, while resilient desert flora, like kingcup cactus and rubber rabbitbrush, and fauna, such as roadrunners and coyotes, thrive in this seemingly harsh environment.

www.nps.gov/whsa/index.htm

THE AMERICAS UNITED STATES (ALASKA)

53 WRANGELL-ST. ELIAS NATIONAL PARK

TO VISIT
BEFORE YOU DIE
BECAUSE

This little-visited national park in south central Alaska is the largest in the United States and offers plenty of opportunities for adventure.

It's hard to talk about Wrangell-St. Elias National Park without using superlatives. Covering over 13 million acres, it's the largest national park in the United States, and in that vast area are four major mountain ranges, the second and third-highest peaks in North America, and the nation's largest glacial system. That's all to say: this is a park with unparalleled opportunities for exploration for hardy adventurers. The most accessible activity? Visiting the Kennecott Mines. From 1911 to 1938, this mining camp pumped out copper around the clock. But when the volume of the high-grade ore diminished, Kennecott became a ghost town. For 60 years, the mill sat abandoned until 1998, when it was purchased by the National Park Service. Now, visitors can tour the 14-story mill, which is still full of machinery, most of which is more than 100 years old. Another good activity? Hiking to Root Glacier. It's an easy four-mile round-trip trek from Kennecott to the face of the 5,000-square-mile ice flow, which is dappled with otherworldly blue pools and streams.

www.nps.gov/wrst/index.htm

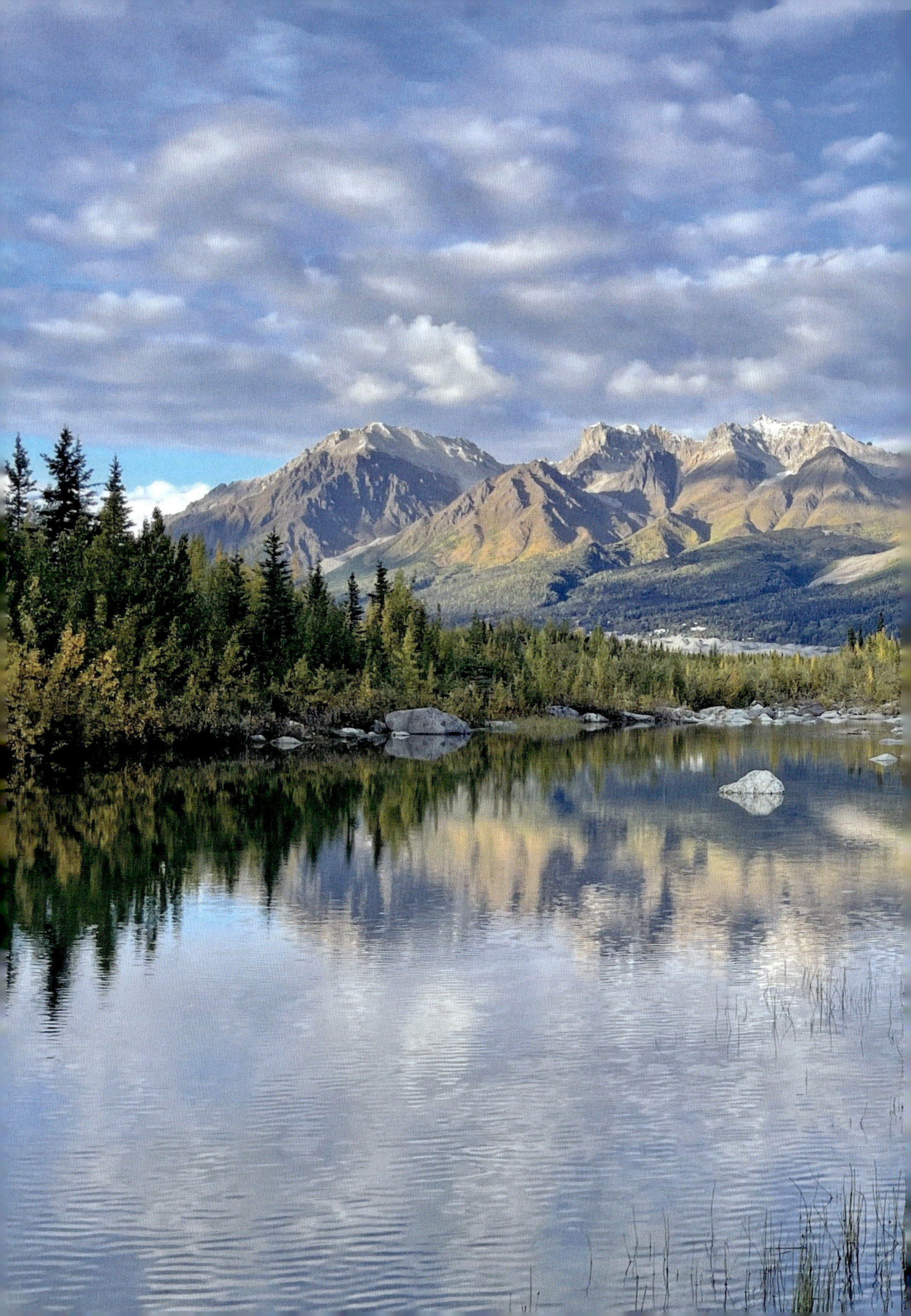

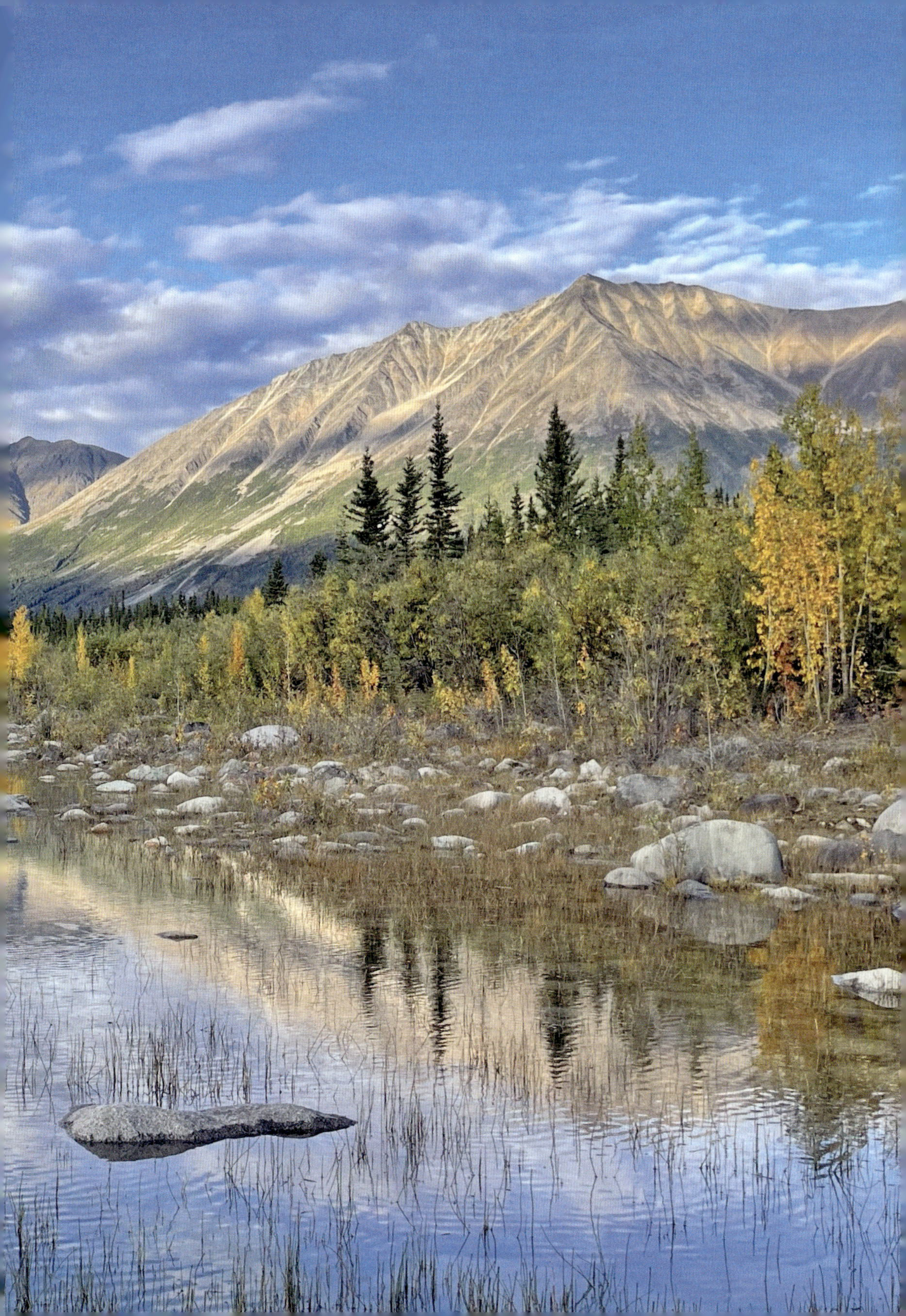

THE AMERICAS

UNITED STATES
(IDAHO/MONTANA/WYOMING)

54 YELLOWSTONE NATIONAL PARK

TO VISIT BEFORE YOU DIE BECAUSE

Marvel at the unique hydrothermal and geological features of Yellowstone, which contains about half of the world's active geysers.

When Yellowstone National Park was established in 1872, it became the first national park in the world. It was the beginning of a movement – today, there are more than 6,000 national parks across the globe. Encompassing over two million acres of pristine wilderness, Yellowstone is home to an astonishing variety of geothermal features, including geysers, hot springs, bubbling mud pots, and fumaroles. The most famous of these is Old Faithful, a geyser that erupts with remarkable regularity, sending a fountain of water more than 100 feet in the air for anywhere between 90 seconds and 5 minutes. Beyond the hydrothermal areas, some of the massive park's other ecosystems include lodgepole pine forests, alpine meadows, grasslands, and wetlands. Together they support a vast array of animals, including the largest free-roaming, wild herds of bison in the U.S., one of the largest elk herds in North America, some of the only grizzly bears in the contiguous U.S., and even some wolverines, lynx, and bighorn sheep.

www.nps.gov/yell/index.htm

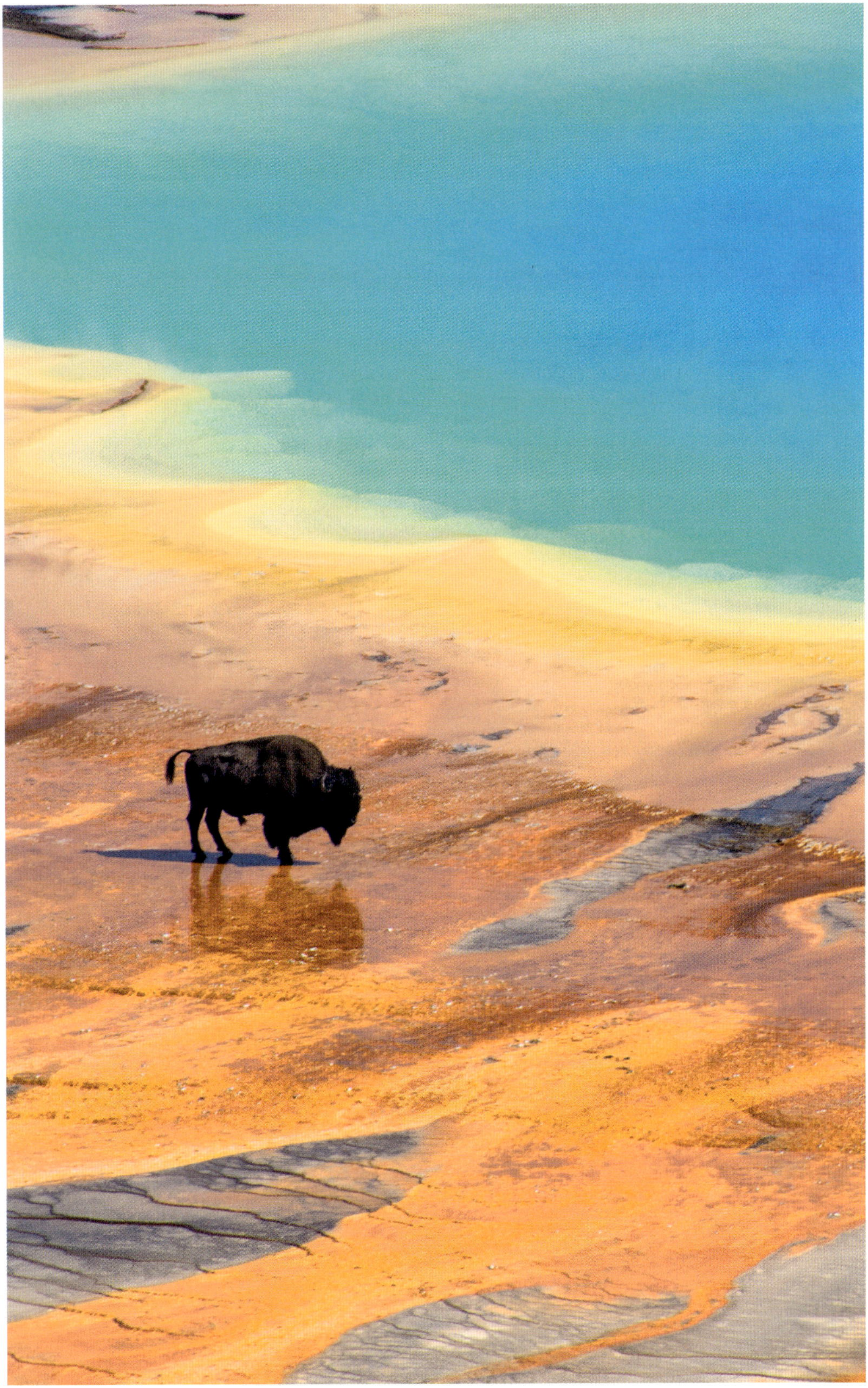

THE AMERICAS   UNITED STATES (CALIFORNIA)

55 YOSEMITE NATIONAL PARK

TO VISIT BEFORE YOU DIE BECAUSE

A shrine to the power of glaciers, the beauty of waterfall's and the tranquility of the High Sierra, it's the park that started the conservation movement.

Though Yosemite wasn't the first national park, it did make history when, in 1864, President Abraham Lincoln signed a congressional bill that set aside the land as a wilderness preserve, something that had been done nowhere else. At the heart of Yosemite National Park lies the Yosemite Valley, carved by ancient glaciers and framed by towering granite monoliths such as El Capitan and Half Dome. These sheer rock formations, rising thousands of feet above the valley floor, are a mecca for rock climbers and adventurers seeking the ultimate challenge. Some of Yosemite's other most famous attractions are its spectacular waterfalls, including the iconic Yosemite Falls, which plunges a staggering 2,425 feet in three breathtaking tiers, and Horsetail Fall, known for its fiery appearance when the setting sun hits it just right in February. The park is also home to abundant wildlife, including black bears, deer, bobcats, coyotes, and over 250 species of birds.

www.nps.gov/yose/index.htm

THE AMERICAS UNITED STATES (UTAH)

56 ZION NATIONAL PARK

TO VISIT BEFORE YOU DIE BECAUSE

Utah's first national park is a geology lover's dream.

Carved over millions of years by the relentless flow of the Virgin River, Zion's towering sandstone cliffs, narrow slot canyons, and high plateaus form a landscape of unparalleled beauty. The park's centerpiece is Zion Canyon, a deep gorge that stretches for miles, its walls rising thousands of feet above the valley floor. The only way to travel the 6.6-mile road that cuts through it is on one of the park's free shuttle buses. But along the way, you can hop off at any of the nine pick-up and drop-off spots, including the head of the Court of the Patriarchs Trail and Zion Lodge, which has quick access to the Emerald Pools Trail, the Grotto Trail, and the Sand Bench Trail. Be sure to keep an eye out for the California condor – the endangered raptor is the largest flying bird in North America. It's one of nearly 300 species of birds found within the park – others include the Mexican spotted owl and the southwestern willow flycatcher.

THE AMERICAS   UNITED STATES (ARIZONA)

57 PETRIFIED FOREST NATIONAL PARK

TO VISIT BEFORE YOU DIE BECAUSE

Gaze up at 225 million years of history preserved in the Painted Desert's colorful sandstone cliffs and petrified prehistoric logs.

Millions of years ago, before the supercontinent Pangaea broke apart, the area that is now Petrified Forest National Park was a lush, tropical forest near the equator. Over time, catastrophic events such as volcanic eruptions and massive floods buried the forest under layers of sediment. Through a remarkable process called petrification, the organic matter in the wood was replaced by minerals such as silica, which, over the course of many thousands of years, crystallized into quartz. During the uplifting of the Colorado Plateau (starting about 60 million years ago), the petrified trees were pushed up to the surface. That movement caused the petrified trees to fracture, giving the appearance of logs cut with a chainsaw and allowing us to see multicolored cross sections of the petrified wood. Another notable part of the park is the Painted Desert area. These colorful badlands, with rocks ranging from deep purples and reds to rich oranges and yellows, are the results of a combination of geological processes, mineral deposits, and environmental conditions. It's an area that's particularly pretty at sunrise and sunset. Hiking trails wind through the otherworldly terrain, leading adventurers past towering hoodoos, massive petrified logs, and remnants of ancient civilizations.

www.nps.gov/pefo/index.htm

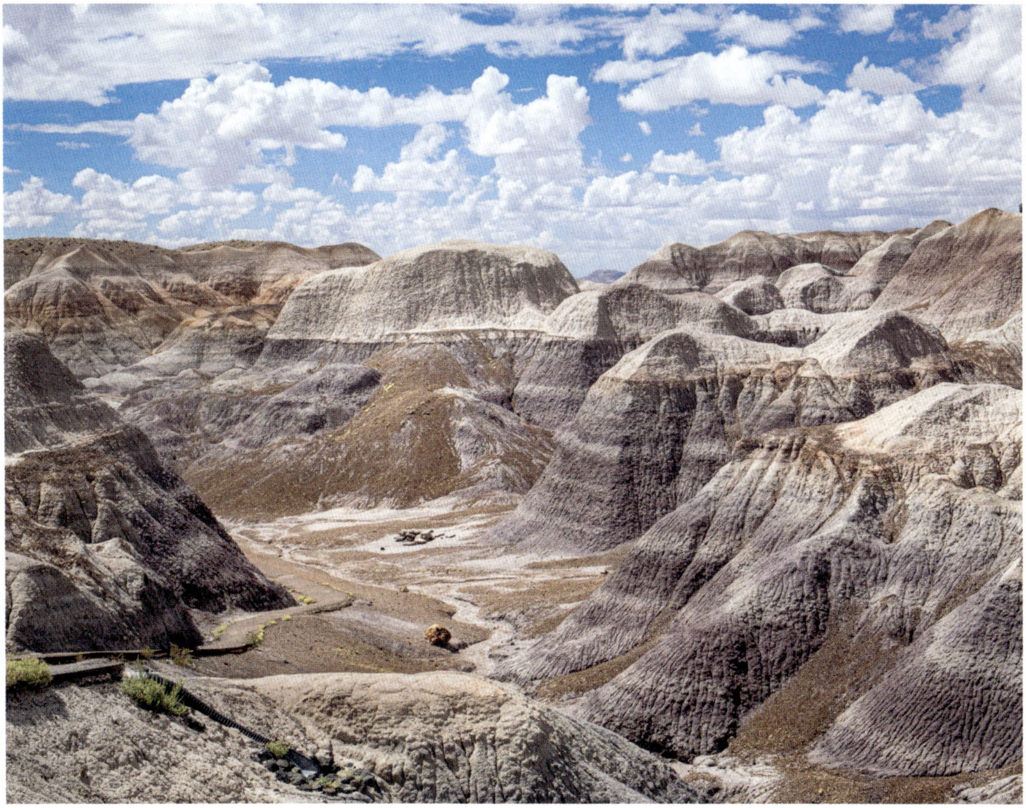

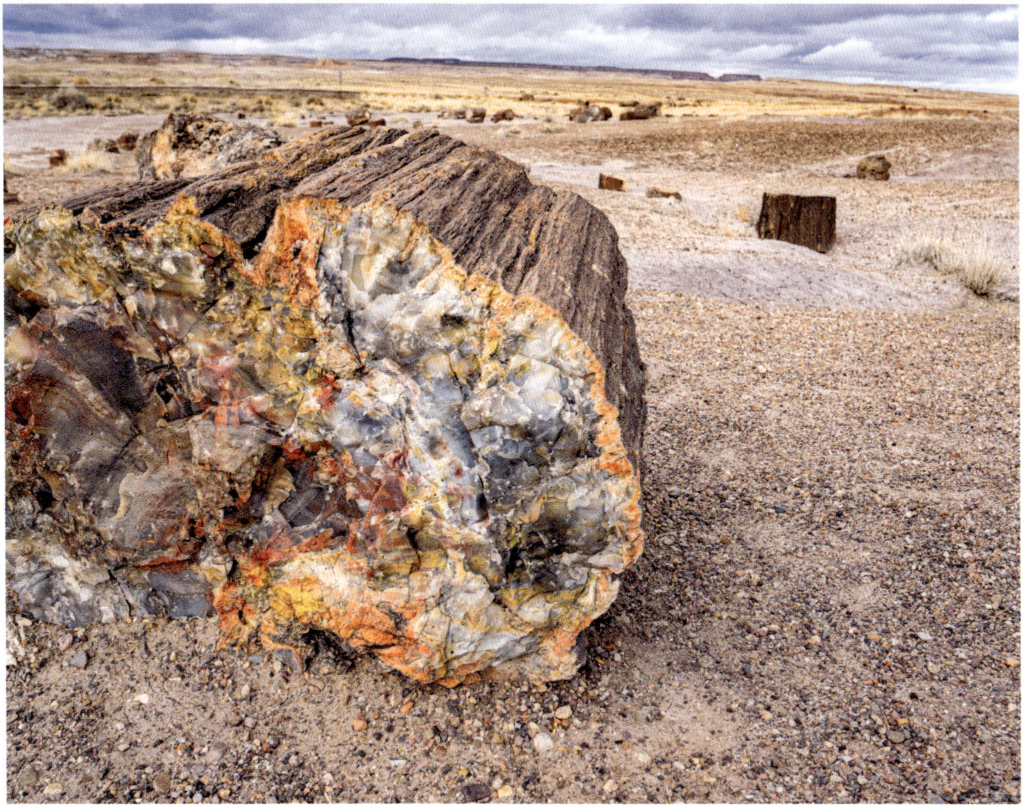

THE AMERICAS UNITED STATES (CALIFORNIA)

58 JOSHUA TREE NATIONAL PARK

TO VISIT BEFORE YOU DIE BECAUSE

Two distinct desert ecosystems, the Mojave and Colorado, intersect in this fascinating landscape sculpted by strong winds.

Arguably, the most distinctive plant in the Mojave Desert grows in this slice of southeastern California. The Joshua Tree, scientifically known as *Yucca brevifolia*, belongs to the agave family. These trees are characterized by their unusual appearance – with twisted bodies and tall, spiky leaves clustered at the ends of thick, spiny branches, they look like something out of a Dr. Seuss book. They can grow to heights of up to 40 feet and have a lifespan of several hundred years. Though odd, Joshua Trees have long played a crucial role in the desert ecosystem. They provide habitat and food for various animals, including desert rodents, insects, and birds. The local Indigenous population once used the tough leaves to weave baskets and sandals, and roasted the seeds to add to meals. Beyond its namesake tree, Joshua Tree National Park is also known for its rugged rock formations, stark desert landscapes, and wide-open skies.

www.nps.gov/jotr/index.htm

THE AMERICAS   UNITED STATES (MONTANA)

59 GLACIER NATIONAL PARK

TO VISIT BEFORE YOU DIE BECAUSE

See what glacial retreat can do to a landscape in this adventurers' paradise dotted with carved valleys, spectacular lakes, and alpine meadows.

Famed conservationist John Muir once called Glacier National Park "the best care-killing scenery on the continent." That's because Glacier National Park encompasses more than a million acres of some of the United States' most pristine wilderness, with alpine meadows, towering peaks, coniferous forests, sparkling lakes, and dozens of waterfalls within its boundaries. Established in 1910, the park is named for the glacially carved landscape. However, about two dozen smaller mountain glaciers from the Little Ice Age are still found in the high country. Here, visitors can explore more than 700 miles of hiking trails, fish for trout in Lake McDonald, or attend a ranger-led program, such as a guided walk or astronomy hour. However, no trip to Glacier National Park is complete without driving along the east-to-west Going-to-the-Sun Road. It spans 50 miles and boasts some of the most jaw-dropping vistas in the park. While you drive, keep an eye out for some of the grizzly bears, mountain goats, bighorn sheep, and wolves found in Glacier.

THE AMERICAS UNITED STATES (ARIZONA)

60 GRAND CANYON NATIONAL PARK

TO VISIT BEFORE YOU DIE BECAUSE

Witness what President Theodore Roosevelt famously said was "the one great sight every American should see."

The Grand Canyon is big. We're talking 277 river miles long, roughly 18 miles from rim to rim at its widest point, and 4,800 feet down from the South Rim (or 6,000 feet from the North Rim). The canyon's sheer scale is humbling, leaving visitors in awe of its vastness and grandeur. The Colorado River carved the Grand Canyon over the course of six million years. The kaleidoscopic colors and unique rock formations help explain the power of erosion and tell the story of Earth's history. The park offers myriad opportunities for exploration and adventure, with numerous hiking trails, scenic overlooks, and adrenaline-pumping activities such as rafting along the river or taking a helicopter tour over the canyon. For those seeking a deeper understanding of the canyon's natural and cultural significance, the park offers informative visitor centers, ranger-led programs, and educational exhibits that delve into its geological history, Native American heritage, and conservation efforts.

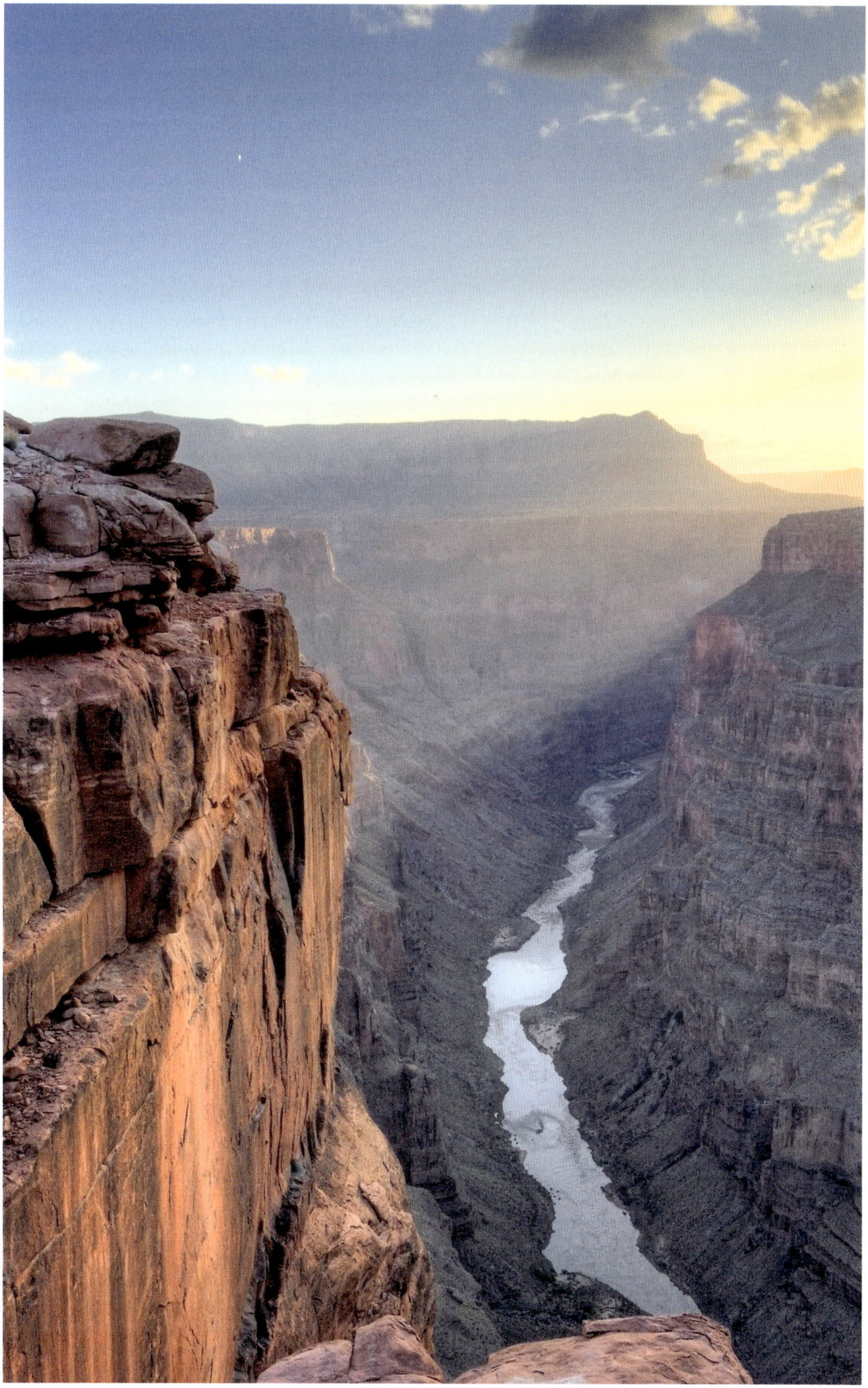

THE AMERICAS   UNITED STATES (ALASKA)

61 KATMAI NATIONAL PARK

TO VISIT BEFORE YOU DIE BECAUSE

Surrounding Novarupta and the Valley of Ten Thousand Smokes, this volcanic area is one of the best brown bear-viewing locations in the world.

More than 2,200 brown bears call the 4-million-acre Katmai National Park home, making it one of the most densely populated areas for the apex predator worldwide. You're virtually guaranteed to see the ursids if you come in the summer, especially if you visit Brooks Falls. The 6-foot-tall, 250-foot-wide waterfall on the Brooks River is a popular fishing spot for the bears – in peak salmon season, as many as 50 bears may perch on the lip of the falls. They're waiting for the silvery fish, who are swimming upstream to their spawning grounds, to launch themselves out of the riverway and into their open jaws. But Katmai offers more than just bear-watching opportunities – there are also oodles of opportunities for adventures, like trekking to Mount Katmai, a large active stratovolcano, or kayaking the 80-mile Savonoski Loop, a wilderness water trail. Another point of interest is the geological wonder of the Valley of Ten Thousand Smokes. After the Novarupta volcano erupted in 1912, the surrounding area was blanketed in ash, in some places to depths of 700 feet. Now, the landscape looks post-apocalyptic, with smoking valleys, steam vents, and lava flows. NASA's Apollo astronauts used the site in 1965 and 1966 because it was believed to be a good representation of the moon.

www.nps.gov/katm/index.htm

THE AMERICAS UNITED STATES (ALASKA)

62 KENAI FJORDS NATIONAL PARK

TO VISIT BEFORE YOU DIE BECAUSE

See dizzying fjords, calving glaciers, and playful humpback whales at the edge of the Kenai Peninsula, a land where the Ice Age lingers.

At the heart of Kenai Fjords are its glaciers, remnants of the last Ice Age that sculpted the land. In fact, more than 50 percent of Kenai Fjords is covered in glacial ice, all stemming from the 700-square-mile, 4000-feet-deep Harding Icefield. It feeds more than 30 floes, including the famous Exit Glacier, which is easily accessible by road and offers visitors a chance to witness the power of glacial retreat up close. Other notable glaciers, such as the Holgate, Aialik, and Pedersen Glaciers, can be explored by boat tours, providing awe-inspiring views of calving ice of all sizes into the sea. The day cruises are the most popular (and easiest) way for visitors to explore the park. During the voyage, guests can scan the rocky coastline for seabirds (nearly 200 species of birds call the park home) and watch humpback whales seem to defy gravity as they breach in Resurrection Bay.

www.nps.gov/kefj/index.htm

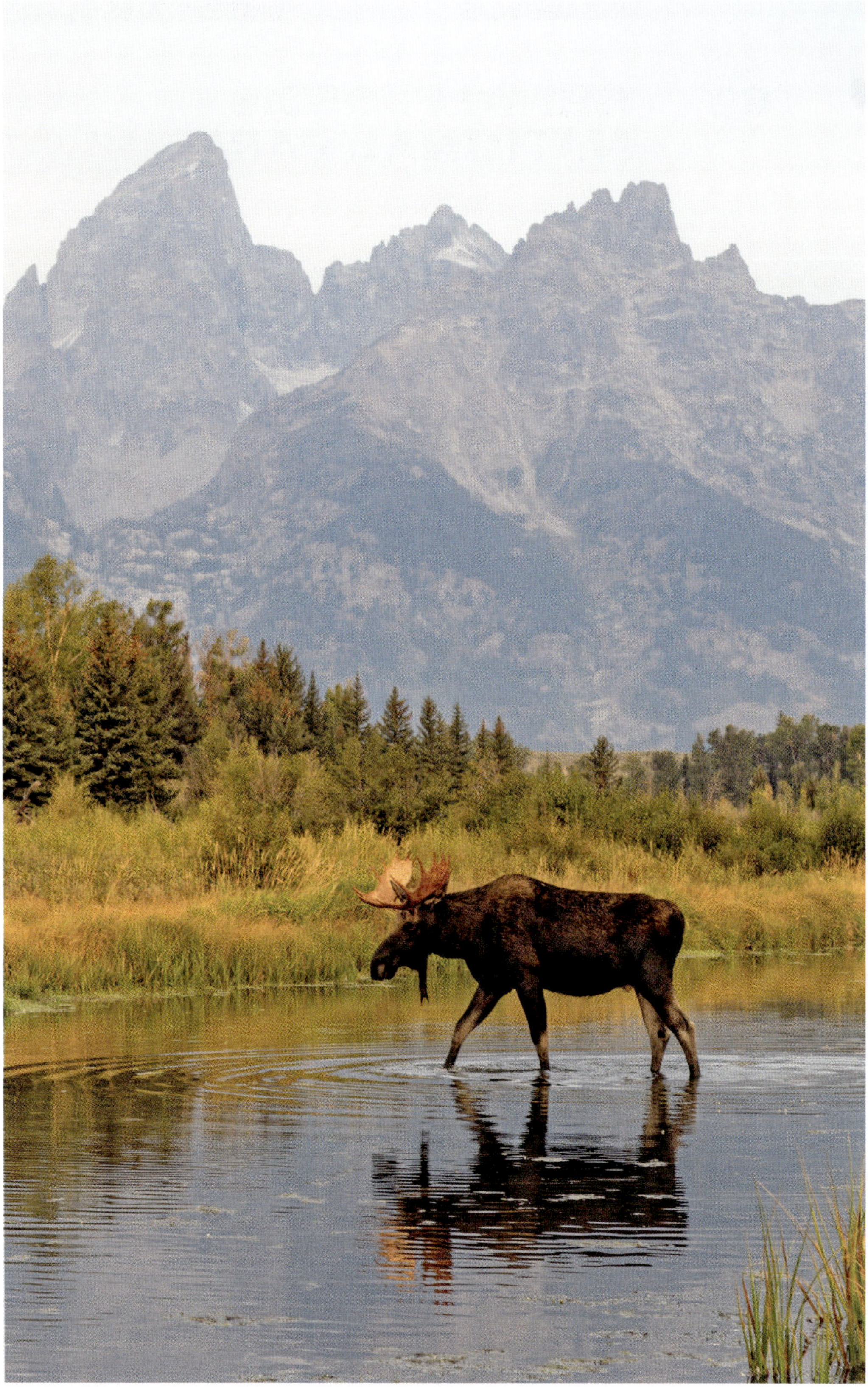

THE AMERICAS UNITED STATES (WYOMING)

63 GRAND TETON NATIONAL PARK

TO VISIT BEFORE YOU DIE BECAUSE

Get a feel for the real American West, in a valley where communities have thrived for over 11,000 years.

Rising 7,000 feet practically straight up from the valley floor, with no foothills to block the views, the sky-piercing serrated peaks of the Grand Tetons are nothing short of dramatic. Visitors to Grand Teton National Park can explore a wealth of accessible outdoor adventures year-round. In the warmer months, hiking trails wind through alpine meadows ablaze with wildflowers while climbers test their skills on the rugged granite faces of the Teton Range. Meanwhile, the Snake River offers world-class fly-fishing opportunities and leisurely rafting, and scenic drives provide access to breathtaking vistas at every turn. As the seasons shift and snow blankets the landscape, Grand Teton and the neighboring resort of Jackson Hole transforms into a winter wonderland. Cross-country skiing, backcountry skiing, and snowshoeing opportunities abound, offering visitors a chance to experience the park in a new light. And throughout the year, visitors can look for wildlife that includes grizzly and black bears, bison, moose, elk, bighorn sheep, and pronghorns.

www.nps.gov/grte/index.htm

THE AMERICAS

UNITED STATES
(TENNESSEE/NORTH CAROLINA)

64 GREAT SMOKY MOUNTAINS NATIONAL PARK

TO VISIT BEFORE YOU DIE BECAUSE

Seek solace in nature or gain a deeper understanding of Appalachian history and culture in the country's most popular national park.

Great Smoky Mountains National Park, nestled on the border between North Carolina and Tennessee in the Appalachian Mountains, is consistently the most-visited national park in the United States. One of the most notable features of the park is its ethereal mist, created by the dense vegetation and abundant water sources, which gave rise to its name. The haze is enchanting, especially in the early morning hours, when it winds its way through the verdant old-growth forests and around the base of the blue-hued mountains. The area also hosts unparalleled biodiversity – nowhere else outside of the tropics has the same level of animal and plant life. For outdoor enthusiasts, the park offers a plethora of recreational opportunities. More than 800 miles of tree-lined hiking trails wind through the wilderness, leading to panoramic vistas, hidden waterfalls, and tranquil streams. The Appalachian Trail, one of the most iconic long-distance hiking routes in the world, passes through the park.

www.nps.gov/grsm/index.htm

THE AMERICAS — UNITED STATES (HAWAI'I)

65 HALEAKALĀ NATIONAL PARK

TO VISIT BEFORE YOU DIE BECAUSE

To see sun rise over a dramatic volcanic landscape that echoes with stories of ancient and modern Hawaiian culture.

This park's most visited attraction is the massive shield volcano, Haleakalā, which means "House of the Sun" in Hawaiian. According to legend, the demigod Maui captured the sun here, releasing it only when it promised to make the days longer. Rising to over 10,000 feet above sea level, Haleakalā is one of the world's largest dormant volcanoes. One of the park's most iconic experiences is witnessing the sunrise from the summit of Haleakalā – as dawn breaks over the horizon, the sky transforms into a canvas of vibrant colors. The park is also an important safe haven for Hawai'i's native flora and fauna, which are under threat from invasive species. More than 400 species of plants are native to the island, and more than 300 are found nowhere else, including the bird-pollinated geraniums, silverswords, and the shrubby na'ena'e. Some of the animals native to the park include the nene goose, the Hawaiian petrel, and the Hawaiian bat.

www.nps.gov/hale/index.htm

THE AMERICAS   UNITED STATES (HAWAI'I)

66 HAWAI'I VOLCANOES NATIONAL PARK

TO VISIT BEFORE YOU DIE BECAUSE

See two of the most active volcanoes in the world in this designated international biosphere reserve and Unesco World Heritage Site.

At the heart of Hawai'i Volcanoes National Park are two of the world's most active volcanoes, Mauna Loa (which last erupted in 1984) and Kilauea (which has been erupting almost continuously for the last 200 years). Kilauea, often called the "drive-in volcano," boasts one of the most accessible volcanic craters on the planet. Visitors can drive along the scenic Crater Rim Drive, stopping at overlooks to witness steaming vents, colorful mineral deposits, and the vast expanse of the Kilauea Caldera. Another of the park's big attractions is the Chain of Craters Road, which descends 3,700 feet from the summit of Kilauea to the coastline. Visitors can witness dormant lava flows, pit craters, and sea arches along the way. The park also offers myriad hiking trails, such as the Thurston Lava Tube Trail, which goes through a cave-like lava tube, and the Pu'uloa Petroglyphs Trail, where visitors can see approximately 23,000 simple etchings carved in stone by Native Hawaiian people.

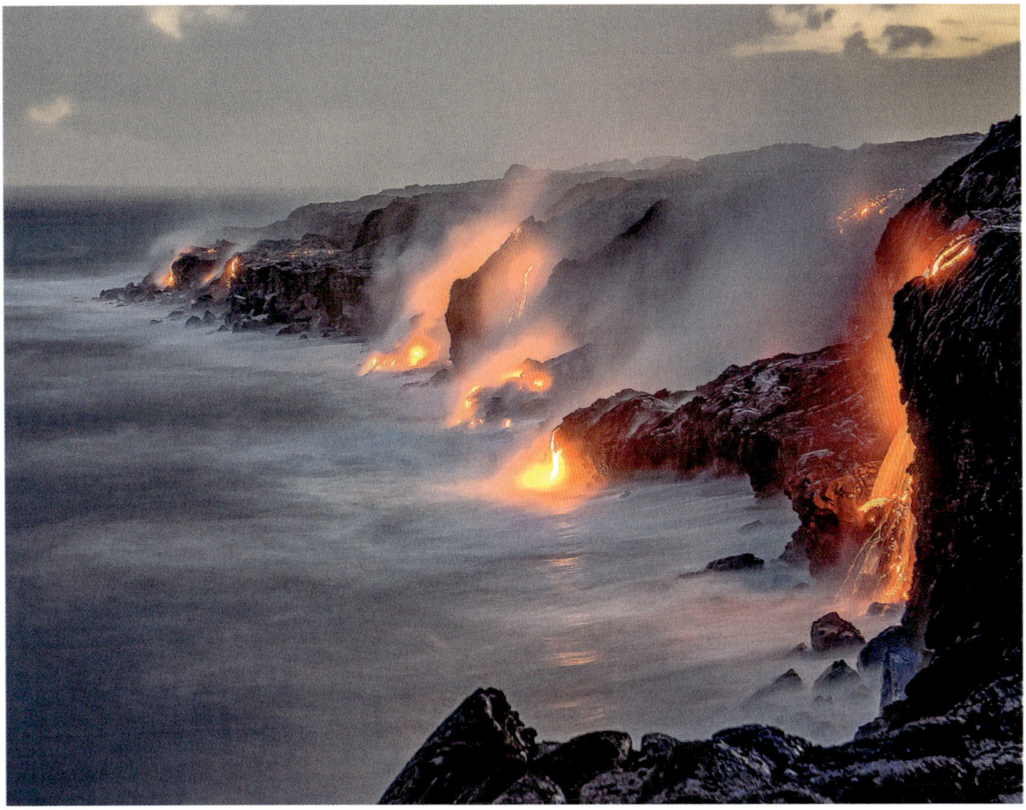

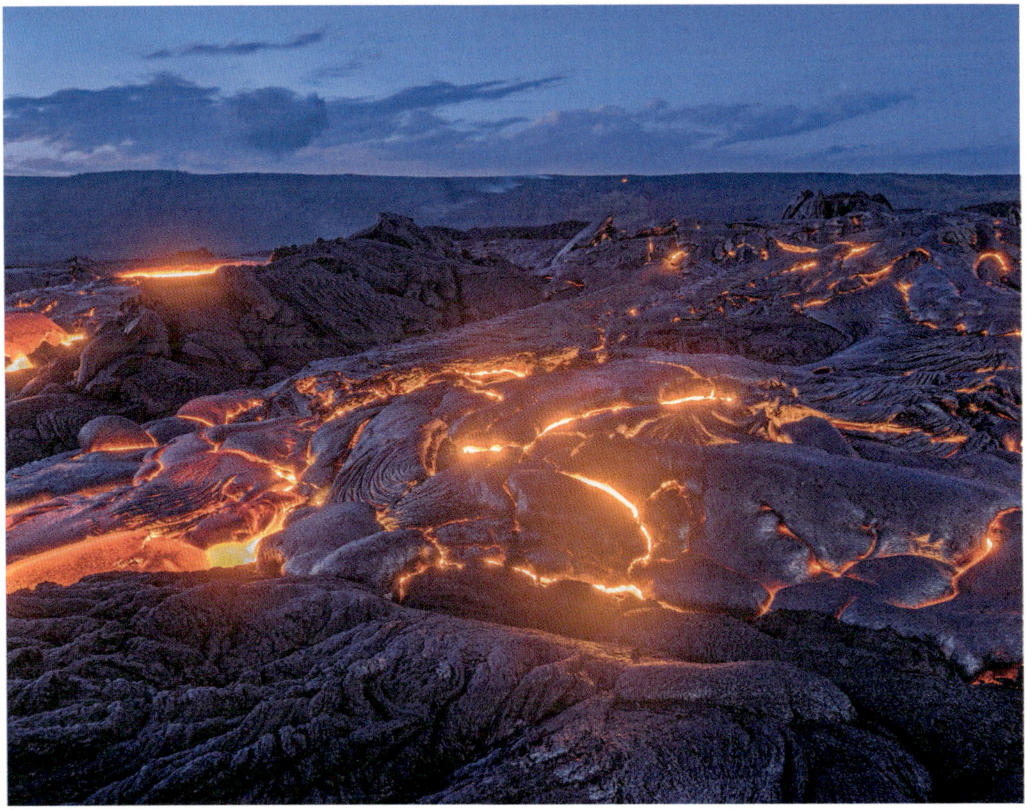

THE AMERICAS UNITED STATES (ALASKA)

67 DENALI NATIONAL PARK

TO VISIT
BEFORE YOU DIE
BECAUSE

Tranquility and wilderness await in this remote park where taiga forests and high alpine tundra culminate in the tallest peak in North America.

Covering over six million acres in the heart of Alaska, the Denali National Park encompasses a diverse range of ecosystems, from boreal forests to alpine tundra, all centered around the towering presence of Denali. Formerly known as Mount McKinley, Denali is North America's tallest peak, rising 20,310 feet above sea level. The mammoth mountain holds cultural significance for Indigenous communities (its name means "The Great One" in Athabascan) and draws thousands of climbers from around the world each year who hope to scale it. While Denali and the other mountains in the Alaska Range are its most famous features, the peaks weren't the reason the park was created. Established in 1917, it was actually the first park in the United States made specifically to protect wildlife, including Alaska's Big Five (bear, moose, Dall sheep, caribou, and wolf). Today, visitors can bike or ride buses along the unpaved 92-mile road into the park to the historic Kantishna mining district, meet with the ranger's sled dog team (it's the only park in the country with a working sled dog kennel), hike along marked trails, or blaze their own.

www.nps.gov/dena/index.htm

THE AMERICAS   UNITED STATES (TEXAS)

68 BIG BEND NATIONAL PARK

TO VISIT BEFORE YOU DIE BECAUSE

Wide-open spaces, desert landscapes, brilliant night skies, and the mighty Rio Grande meet at the end of the road in Far West Texas.

Spanning over 800,000 acres in southwest Texas, Big Bend National Park is sometimes referred to as "three parks in one." That's because it encompasses so many ecosystems, including river areas, deserts, and mountains. One of the biggest draws is the Rio Grande River, which carves its way through deep limestone canyons on the park's southern side, forming the natural border between the United States and Mexico. Rafting or canoeing along the river offers a unique perspective on the rugged beauty of the landscape, and offers opportunities to spot wildlife such as bald eagles, javelinas, and even the occasional black bear. As night falls, the skies above Big Bend come alive with a dazzling display of stars, thanks to the park's remote location far from the glare of city lights. Astronomy enthusiasts flock to the park to marvel at the Milky Way stretching across the heavens while campers gather around crackling bonfires to swap stories beneath the vast expanse of the cosmos.

www.nps.gov/bibe/index.htm

THE AMERICAS UNITED STATES (CALIFORNIA)

69 REDWOOD NATIONAL PARK

TO VISIT BEFORE YOU DIE BECAUSE

Wonder at the tallest living tree in the world and enjoy scenic drives and hikes between the majestic trunks of the ancient giant redwoods.

Redwood trees are the tallest living organisms on Earth, sometimes reaching heights of more than 350 feet. The tallest living tree in the world right now? It's found in a remote area of Redwood National Park in California, stands 379.1 feet, has a diameter of more than 16 feet, and is named Hyperion. It's estimated to be between 600 and 800 years old. The park is also home to the second, fourth, and fifth tallest known trees (currently 377, 371, and 363 feet, respectively). As you step into the park, you're immediately enveloped by the majestic presence of these ancient giants, their massive trunks stretching toward the sky and their branches reaching out like fingers. One of the park's most iconic attractions is the Avenue of the Giants, a scenic drive that winds through a dense concentration of old-growth redwoods, showcasing some of the largest and oldest trees in the world.

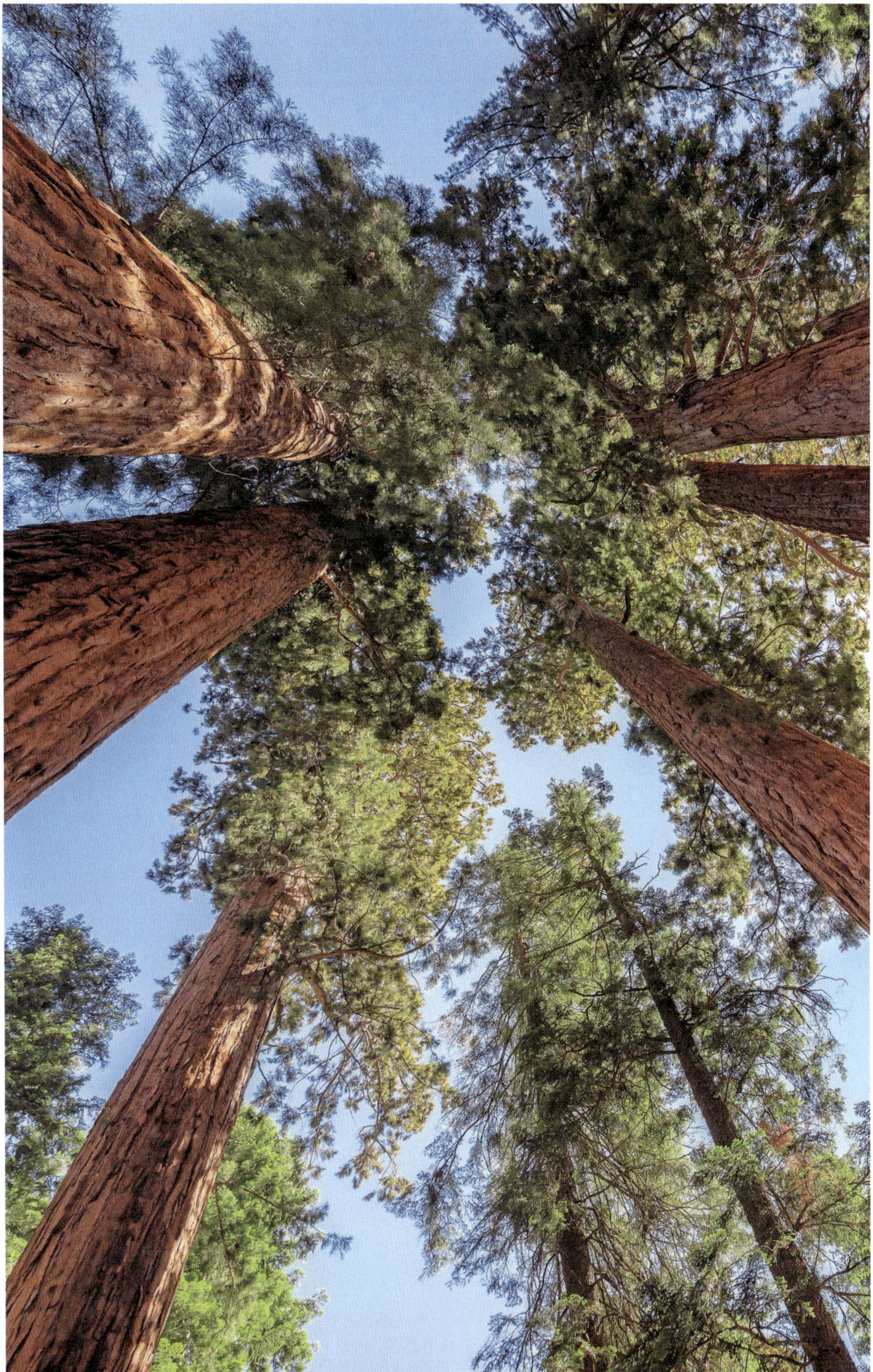

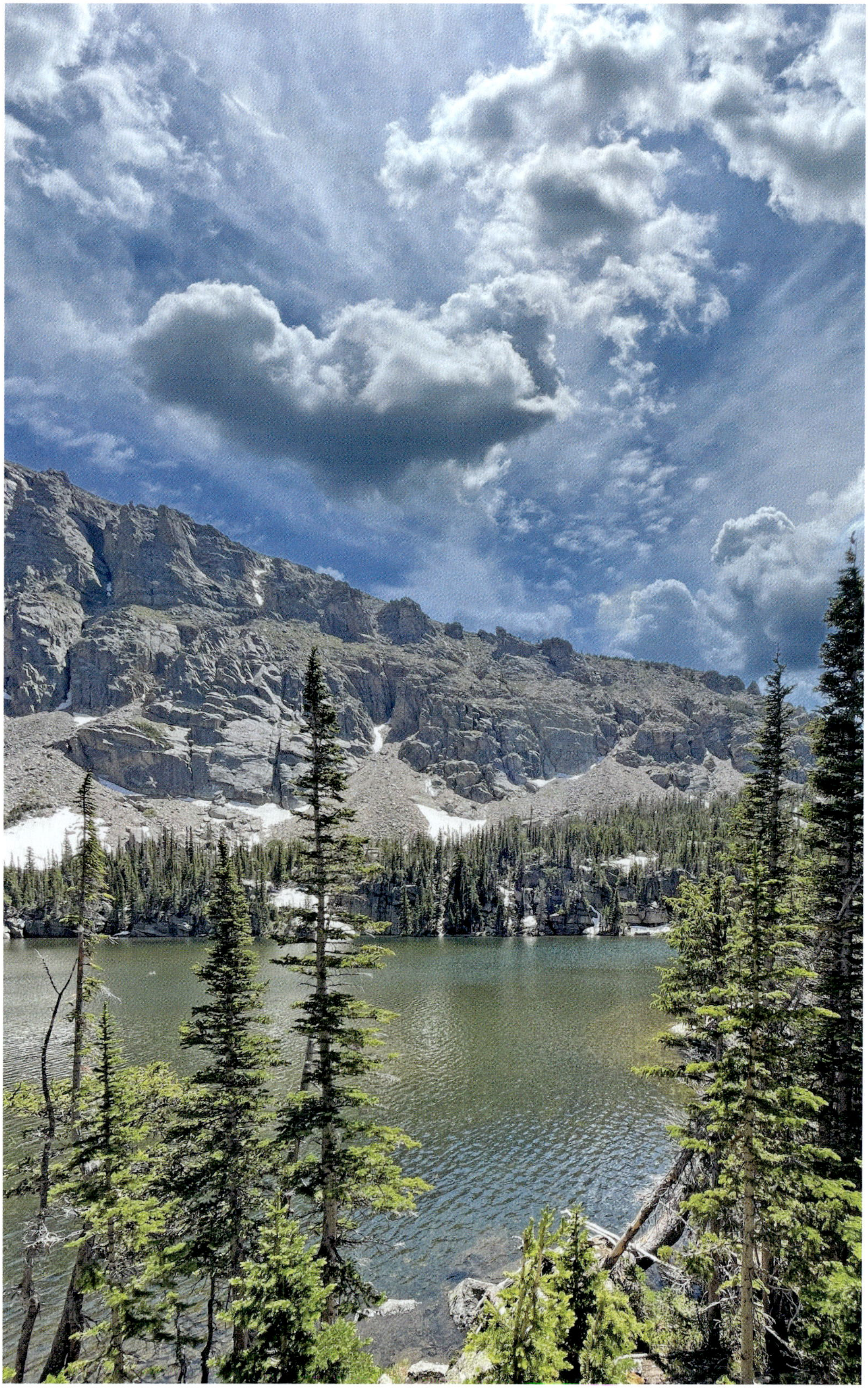

THE AMERICAS UNITED STATES (COLORADO)

70 ROCKY MOUNTAIN NATIONAL PARK

TO VISIT BEFORE YOU DIE BECAUSE

Experience life above the treeline and traverse the Continental Divide at this 415-square-mile park known as a land of extremes.

As the name might suggest, the Rocky Mountains are the main draw of this national park. With over 60 peaks rising above 12,000 feet in elevation, it's an enticing destination for adventurers and mountaineers alike. The tallest mountain is Longs Peak, which stands sentinel at 14,259 feet. The park features hundreds of miles of hiking trails that meander through alpine meadows, dense forests, and rough-and-tumble mountain terrain, catering to hikers of all skill levels. Visitors can embark on day hikes to serene lakes such as Bear Lake or challenge themselves with multi-day backpacking adventures into the park's remote wilderness areas. Scenic drives wind their way through the park, offering awe-inspiring vistas at every turn. Trail Ridge Road, often dubbed "Highway to the Sky," traverses the Continental Divide, providing panoramic views of the surrounding valleys and mountain meadows. The nationally designated All American Road is also the highest continuous paved road in the United States, offering access to alpine tundra ecosystems that are otherwise not easily seen.

www.nps.gov/romo/index.htm

THE AMERICAS · UNITED STATES (CALIFORNIA)

71 SEQUOIA AND KINGS CANYON NATIONAL PARK

TO VISIT BEFORE YOU DIE BECAUSE

The largest tree in the world lives here, in one of the park's 40 giant sequoia groves, set on a rolling plateau on the western slopes of the Sierra Nevada.

These two contiguous parks, often managed jointly, encompass a vast expanse of over 800,000 acres in California, offering visitors a diverse array of natural wonders to explore. At the heart of Sequoia National Park lies the famed Giant Forest, home to some of the largest and oldest trees on Earth, including the General Sherman Tree, which stands as the world's largest living organism by volume (it's 274.9 feet fall and has a ground circumference of 102.6 feet). General Sherman and some of the other giant sequoias in the park are estimated to be somewhere between 2,300 and 2,700 years old. Kings Canyon National Park boasts attractions including the namesake Kings Canyon, a dramatic glacial valley with some of North America's steepest vertical relief. Here, visitors can also marvel at cascading waterfalls, such as the spectacular and powerful Grizzly Falls, Mist Falls, and Roaring River Falls.

www.nps.gov/seki/index.htm

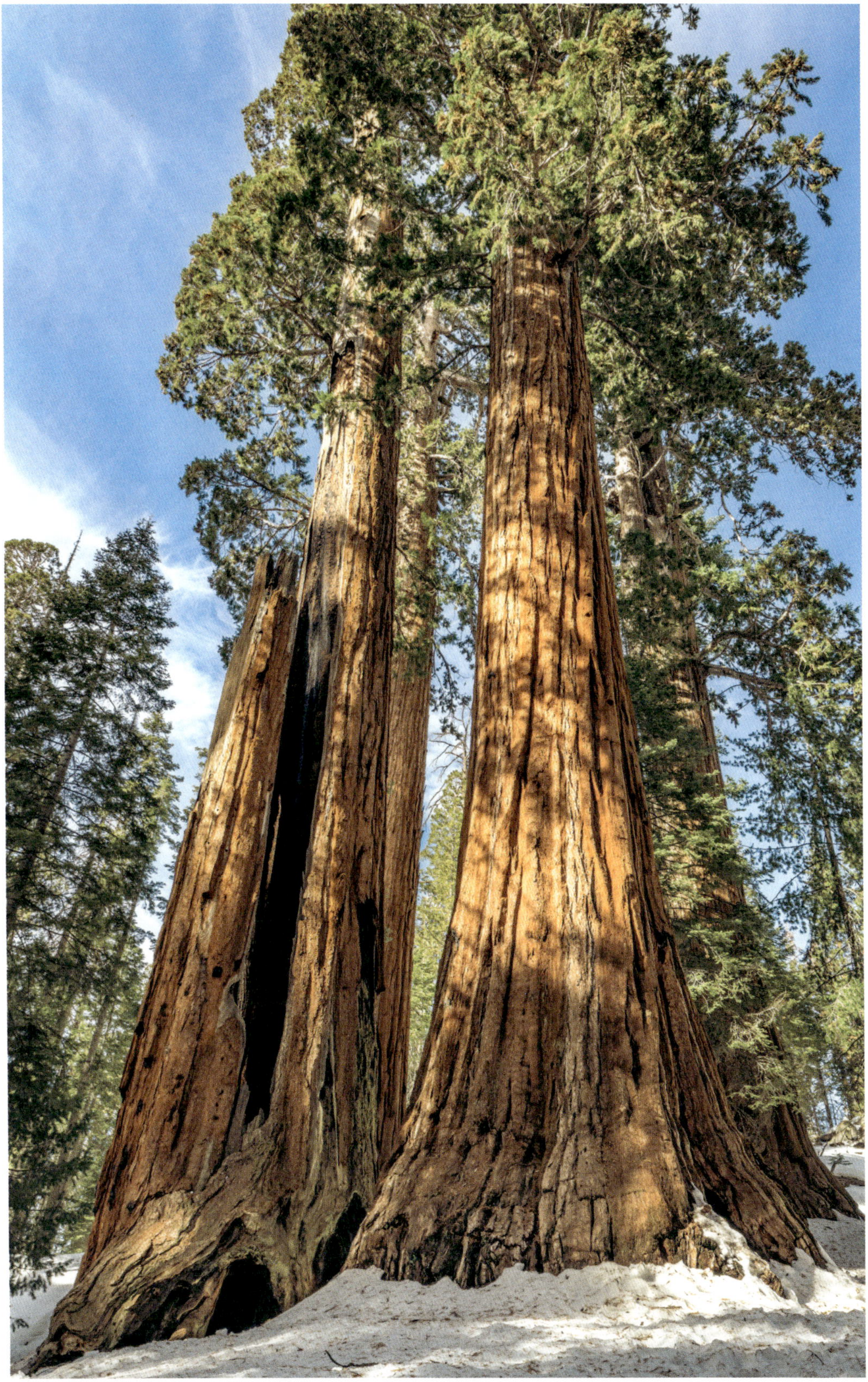

THE AMERICAS   UNITED STATES (VIRGINIA)

72 SHENANDOAH NATIONAL PARK

TO VISIT BEFORE YOU DIE BECAUSE

A short drive from Washington, D.C., this park offers scenic drives and oodles of trails, including part of the Appalachian Trail.

Just 75 miles from the nation's capital, Shenandoah National Park beckons city dwellers with more than 500 miles of trails, including 101 miles of the famed Appalachian National Scenic Trail. The park is particularly popular during the summer when wildflowers and flowering shrubs bloom, and in the fall, when the leaves of the hickories, maples, oaks, and other hardwood trees turn fiery shades of yellow, orange, and red. During that season, the Skyline Drive, a 105-mile-long scenic roadway that runs through this slice of the Appalachian Blue Ridge Mountains, is busy but well worth the effort. These protected lands have many scenic cascades, like Dark Hollow Falls, Doyles River Falls, Hazel River Falls, and Rose River Falls. And even though it's so close to a handful of major cities, there's a diverse wildlife population, including black bears, white-tailed deer, bobcats, big brown bats, migratory birds, and the endangered Shenandoah salamander.

www.nps.gov/shen/index.htm

THE AMERICAS UNITED STATES (UTAH)

73 ARCHES NATIONAL PARK

TO VISIT BEFORE YOU DIE BECAUSE

The fiery reds, oranges, and yellows of the stones contrast beautifully against the deep blue sky in this landscape filled with unique rock formations.

Arches National Park, located in southeastern Utah in the United States, is a haven for everyone from geology nerds to outdoor adventurers. What makes the park unique is its strange and captivating landscape, where wind and water have sculpted an otherworldly array of sandstone arches and towering pinnacles over the course of millions of years. Covering over 76,000 acres of high desert terrain, the park is a testament to the relentless forces of erosion and geologic upheaval. All told, there are more than 2,000 arches in the park, ranging from paper-thin cracks to the Landscape Arch, which spans 290 feet, making it the longest arch in the world. It and other awe-inspiring crimson-hued formations, like the Delicate Arch and the Windows Arch, draw visitors from around the globe each year.

nps.gov/arch/index.htm

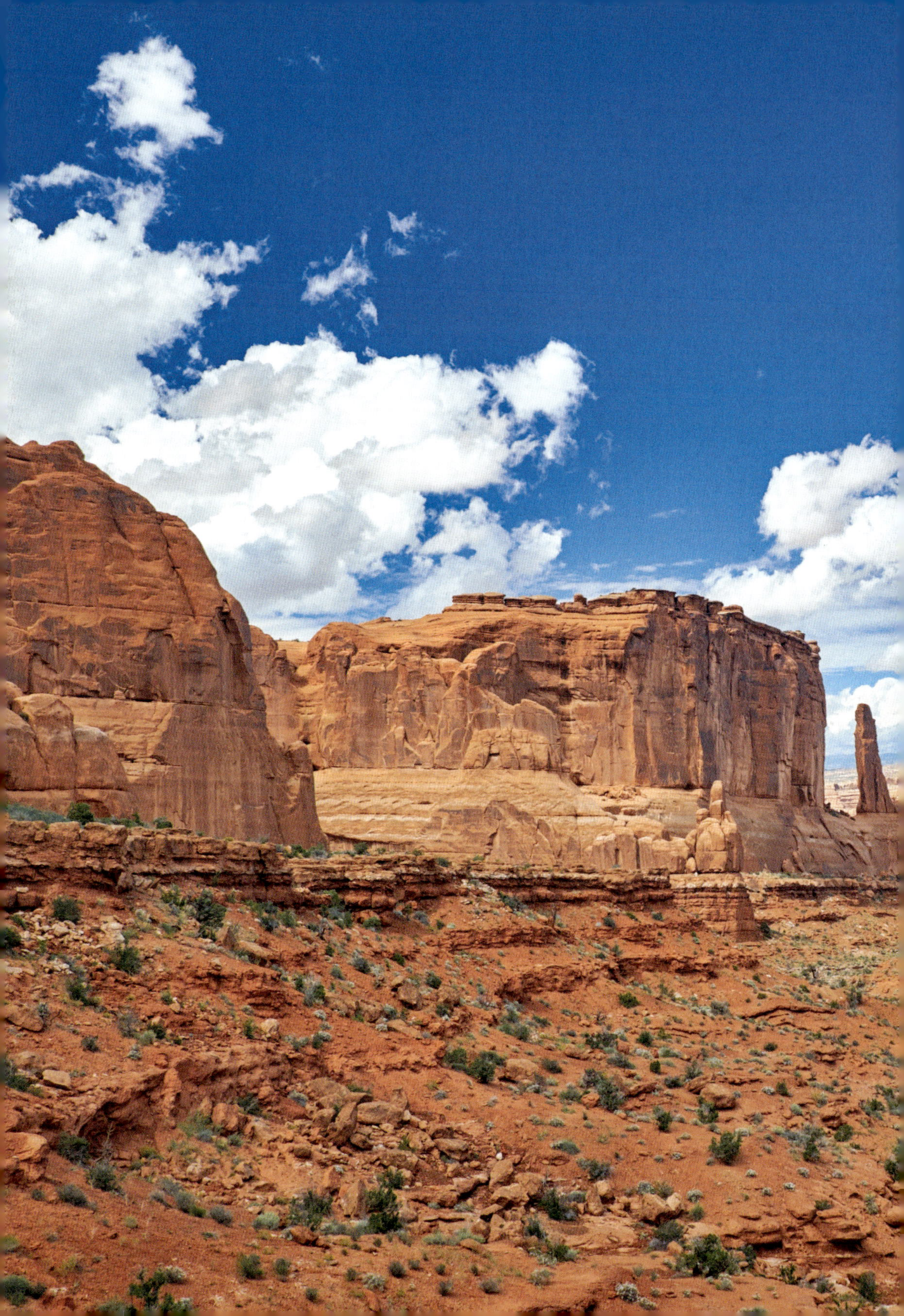

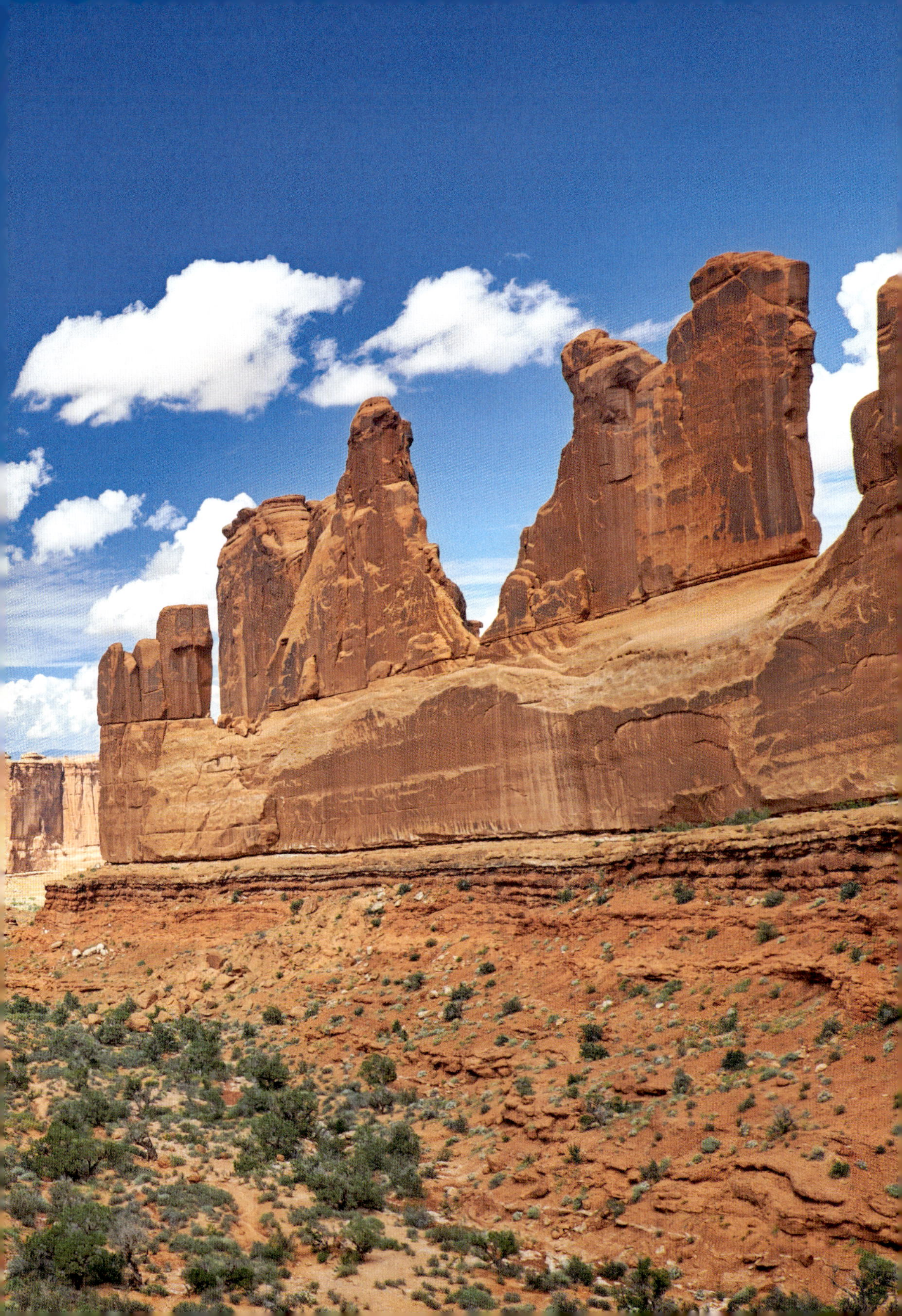

THE AMERICAS UNITED STATES (MAINE)

74 ACADIA NATIONAL PARK

TO VISIT BEFORE YOU DIE BECAUSE

This crown jewel of the North Atlantic coastline, a key site for the Wabanaki, or "People of the Dawnland," is where the northern forests meet the ocean.

Encompassing over 49,000 acres of granite-domed mountains, lush spruce and fir forests, rocky shores, and pristine lakes, Acadia National Park has long drawn travelers with its tapestry of natural wonders. At the heart of the park stands Cadillac Mountain, the tallest peak on the eastern seaboard. Visitors can drive to the top and take in the breathtaking panoramic views of the surrounding Atlantic Ocean and neighboring islands – it's especially spectacular during sunrise and sunset. It's far from the only remarkable geological formation, though. Another awe-inspiring spot to visit is Thunder Hole, with its dramatic sea stacks carved into existence by millennia of coastal erosion. Myriad trails crisscross the park, from coastal walks to hidden coves to leg-burning hikes to see glacial lakes fringed by verdant forests. And because it's a coastal park, maritime activities are plentiful, including kayaking, sailing, and beach combing. It's also possible to visit Bass Harbor Head Lighthouse and Abbe Museum to learn about the region's seafaring past and Wabanaki Nations' heritage.

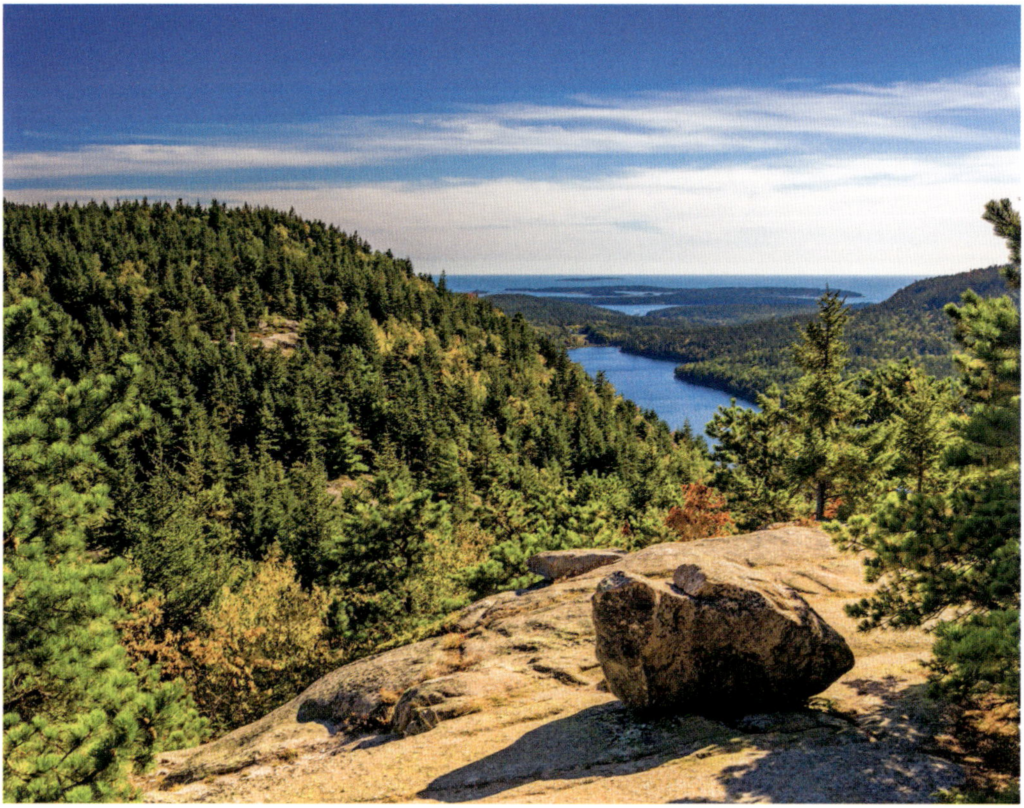

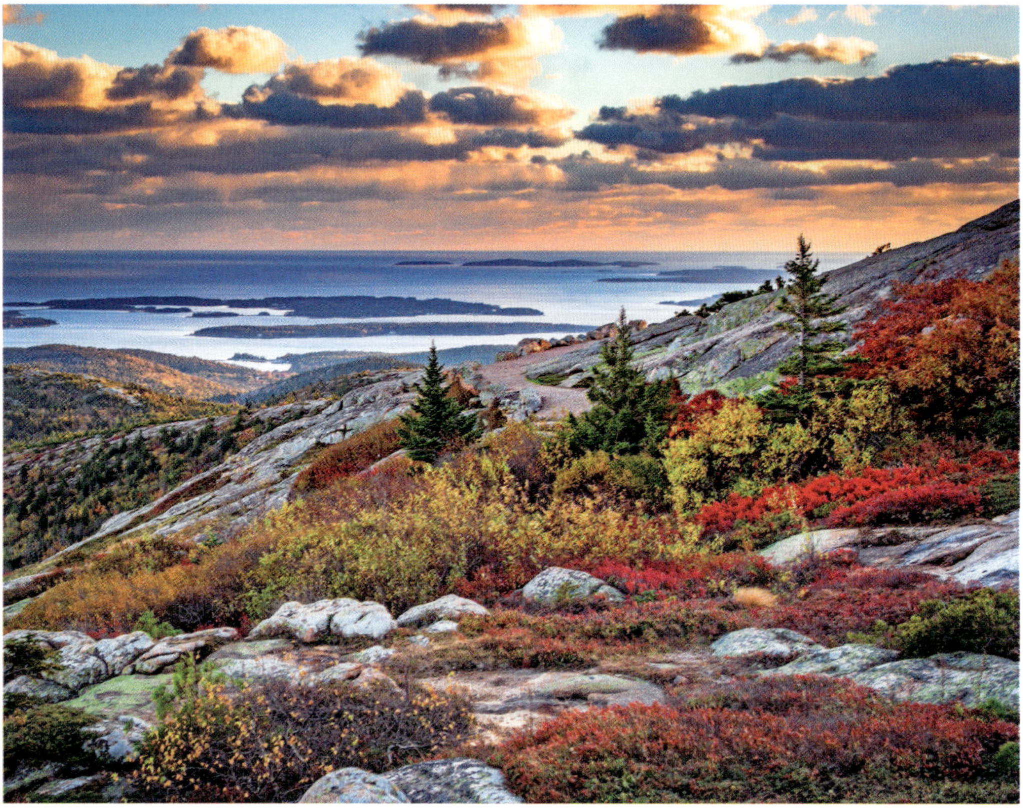

THE AMERICAS UNITED STATES (CALIFORNIA)

75 CHANNEL ISLANDS NATIONAL PARK

TO VISIT BEFORE YOU DIE BECAUSE

This handful of remarkable islands is home to a vast array of animals, many of which can be found nowhere else.

Encompassing five islands (Anacapa, Santa Barbara, Santa Cruz, San Miguel, and Santa Rosa), this national park off the coast of southern California is also known as the "North American Galápagos." That moniker is thanks to the breadth of endemic species that live here – thousands of years of isolation created some very unique animals, including the island fox, the island spotted skunk, and the island deer mouse. Millions of shorebirds, like the black oystercatcher and double-crested cormorant, also use the islands as nesting and feeding grounds. In fact, the Channel Islands support the largest breeding colonies of seabirds in southern California. And because the park boundaries stretch a mile offshore, a wide variety of sea animals are within the protected land. That includes pinnipeds, such as the northern elephant seal, California sea lion, and Guadalupe fur seal, and cetaceans, like the gray whale, humpback whale, and bottlenose dolphin. Beyond wildlife, the islands are known for their sea caves (including the second-largest in North America), springtime wildflowers, and opportunities for hiking and diving.

www.nps.gov/chis/index.htm

THE AMERICAS VENEZUELA

76 CANAIMA NATIONAL PARK

TO VISIT BEFORE YOU DIE BECAUSE

Feel the spray from the world's highest waterfall, Angel Falls, which plunges off the iconic table-top plateaus of this part of southeastern Venezuela.

One of the Canaima National Park's most iconic features is Angel Falls, the world's highest uninterrupted waterfall. Plummeting from the summit of Auyán-tepui, Angel Falls cascades an astonishing 3,211 feet, casting a delicate mist that blankets the surrounding jungle. The sight of this majestic natural wonder is genuinely awe-inspiring, drawing visitors from around the globe to witness its grandeur. Aside from Angel Falls, Canaima National Park boasts a myriad of other enchanting waterfalls, such as Hacha, Wadaima, and Sapito Falls. Another popular reason to visit is to see the mesmerizing Gran Sabana, a vast savannah dotted with dramatic table-top mountains known as tepuis. These towering sandstone plateaus, some rising over 1,000 yards high, create a surreal and otherworldly backdrop against the endless horizon. The park claims a diverse array of flora and fauna, with thousands of plant species, such as orchids and bromelias, and hundreds of bird species, including the harpy eagle and the Guianan cock-of-the-rock.

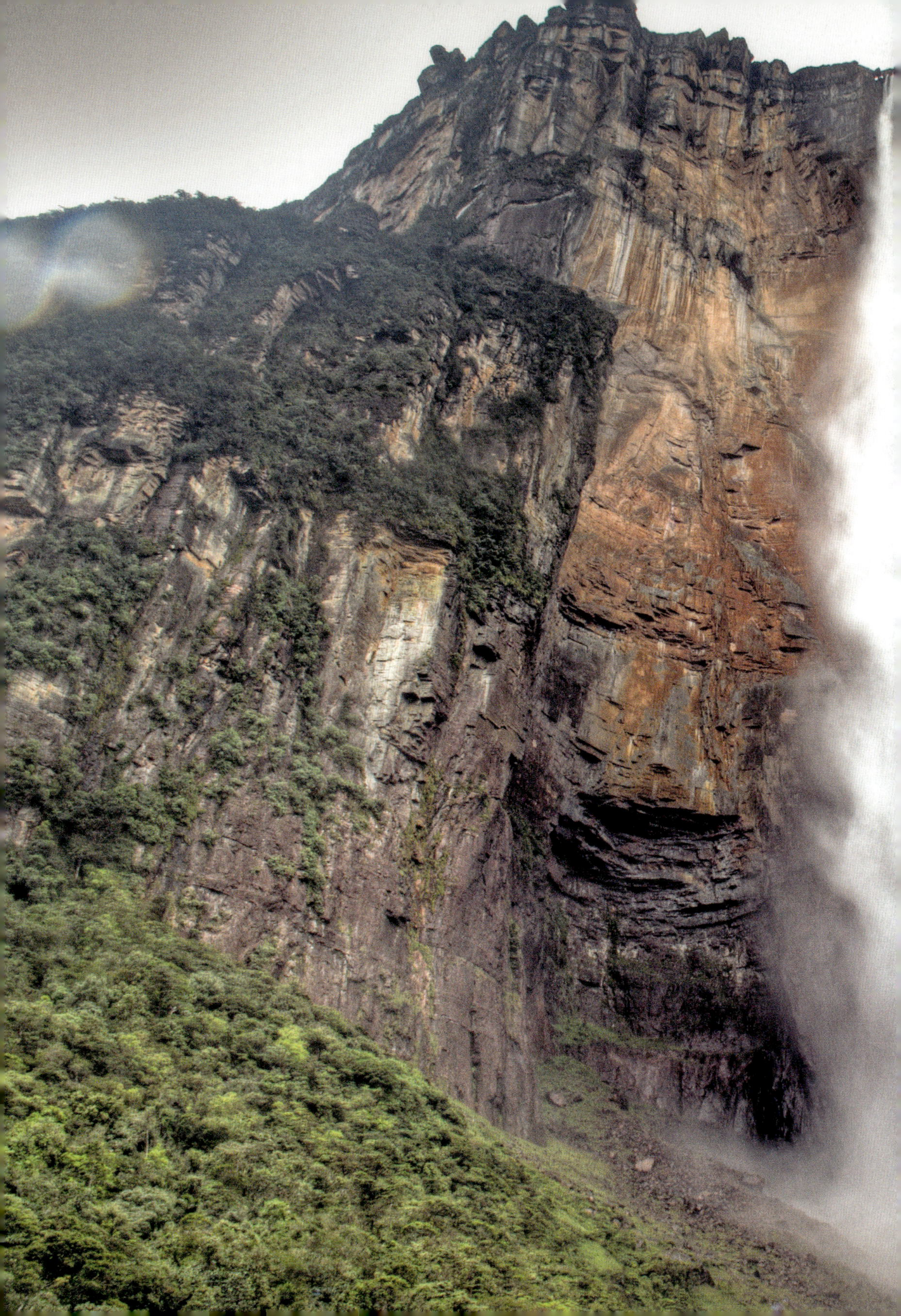

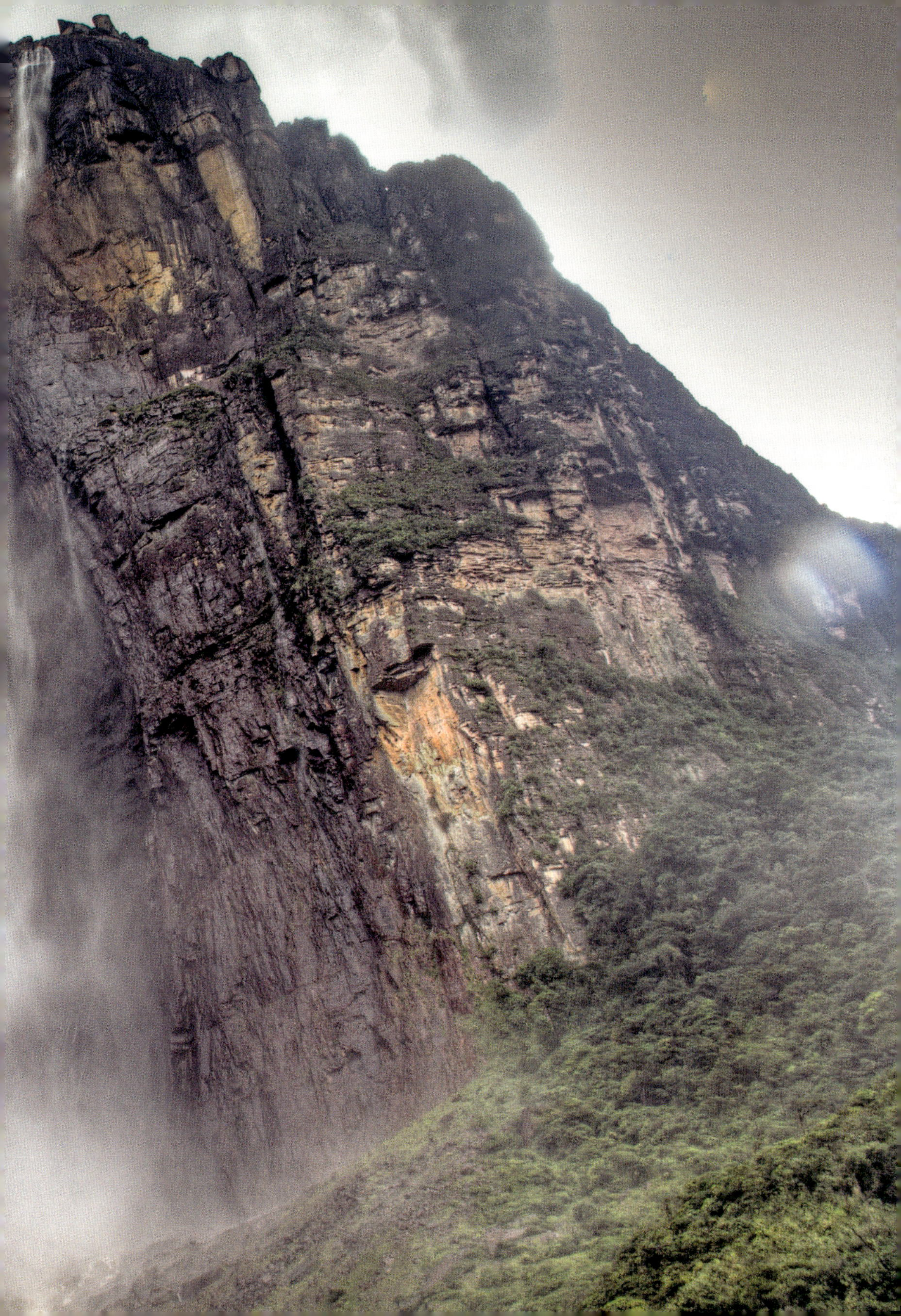

ASIA — BHUTAN

77 JIGME DORJI NATIONAL PARK

TO VISIT BEFORE YOU DIE BECAUSE

This Himalayan paradise offers stunning wildflower meadows and rare animal sightings of endangered species such as the snow leopard.

Established in 1974 to protect the habitat of snow leopards, clouded leopards, and the Bengal tiger's, and named after the third king of Bhutan, Jigme Dorji Wangchuck, this is the second-largest protected area in Bhutan, a country that takes conservation seriously. Here, visitors will find mountainous terrain, milky glacial lakes, coniferous forests, and alpine meadows where wildflowers like the blue poppy (the national flower) are abundant. The highest point is Mount Jomolhari (nicknamed the "Bride of Kangchenjunga"), which reaches 24,035 feet. It's considered sacred to the Tibetan Buddhists, who believe it is home to one of the Five Tsheringma Sisters. Two historical fortresses, Gasa Dzong and Lingshi Dzong, are also within the park boundaries (the latter still houses a few dozen monks). Jigme Dorji National Park is home to many animals you'd be hard-pressed to find elsewhere, including several endangered or vulnerable species such as the Bhutan takin, Himalayan blue sheep, and red panda.

national-parks.org/bhutan/jigme-dorji

ASIA — BORNEO (MALAYSIA)

78 KINABALU NATIONAL PARK

TO VISIT BEFORE YOU DIE BECAUSE

The total number of plant species found in Kinabalu, a 291-square-mile park to the northeast of the island of Borneo, is more than in all of Europe combined.

Named after the 13,435-feet-high Mount Kinabalu, Malaysia's Kinabalu National Park would be a good place to start if you were hoping to see some of Borneo's rarest endemic flora and fauna. Representatives from half of all of Borneo's plant species, half of all of Borneo's mammals, birds, and amphibians, and two-thirds of Borneo's reptiles are found on this protected land. Its rich flora includes thousands of plant species, including more than 1,700 types of orchids alone. There are also pitcher plants, various carnivorous plants, and the Rafflesia, known as the world's largest flower (its giant red blossom can grow over 170 centimeters in diameter). Beyond its botanical treasures, Kinabalu National Park boasts a remarkable diversity of wildlife, including orangutans, gibbons, Oriental small-clawed otters, clouded leopards, Bornean ferret-badgers, three kinds of deer, and over 300 species of birds. In fact, the park is a paradise for birdwatchers, with sightings of vibrant bird species like the Bornean green magpie and the majestic rhinoceros hornbill.

www.sabahparks.org.my/kinabalu-park

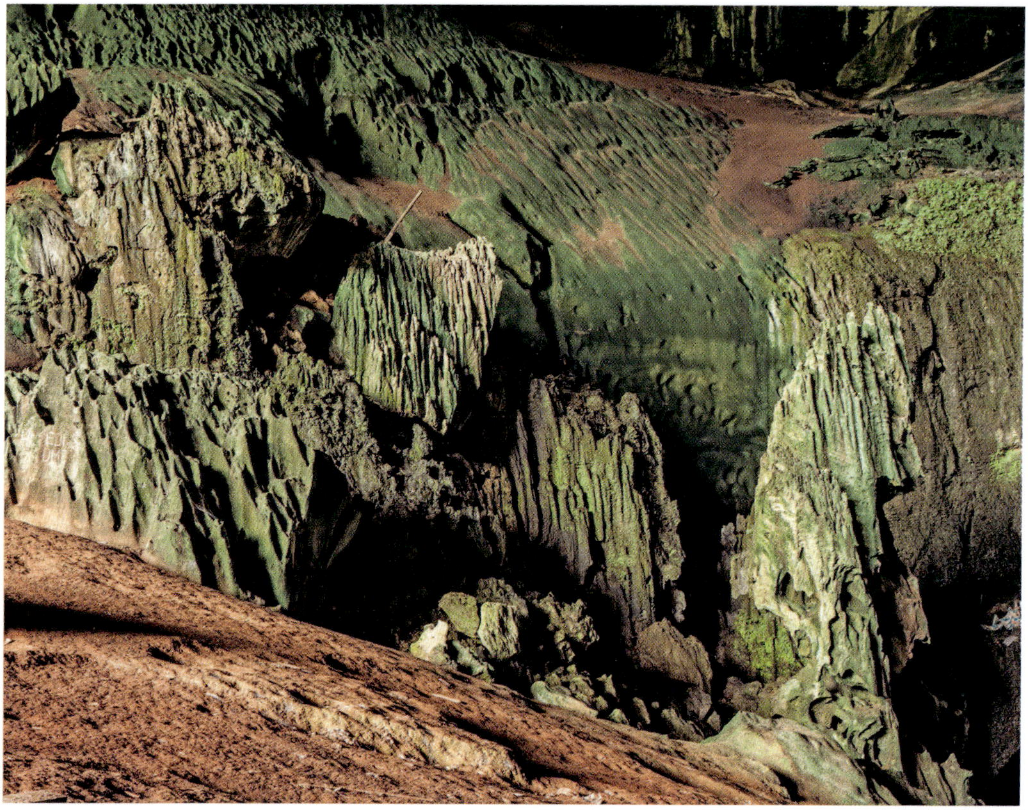

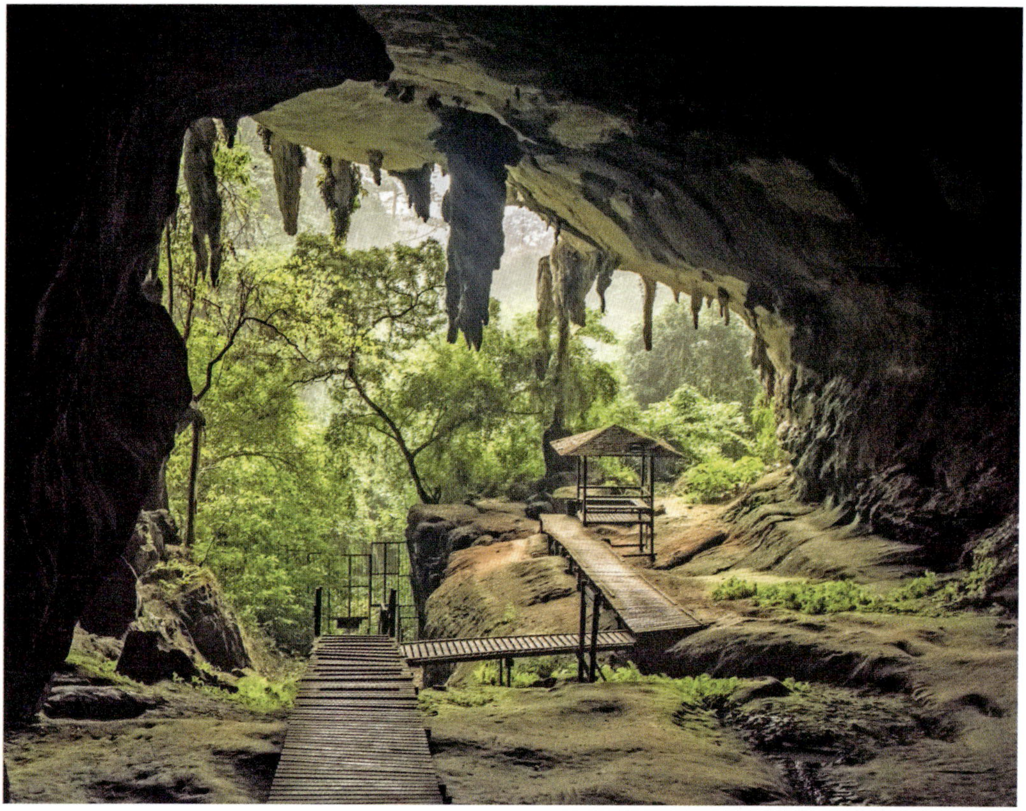

ASIA   BORNEO (MALAYSIA)

79 NIAH NATIONAL PARK

TO VISIT
BEFORE YOU DIE
BECAUSE

This small park in Sarawak, Borneo holds a massive cave system that has been inhabited for more than 40,000 years.

Within Malaysia's Niah National Park lies the Niah Cave Complex, a sprawling network of caves that have played a significant role in human history. The Great Cave, one of the largest limestone caves in the world, with its cathedral-like chambers, boasts towering stalactites and stalagmites, some of which have formed over millions of years. Beyond its geological marvels, Niah National Park holds immense archaeological significance. Similarly, the Painted Cave has rock paintings dating as far back as 1,200 years ago, while other caves have yielded evidence of human habitation dating back as far as 40,000 years, making this one of the most important archaeological sites in Southeast Asia. Excavations have unearthed ancient tools, pottery, and skeletal remains, shedding light on the prehistoric peoples who once called these caves home. Guided tours are available for those eager to explore the caves and learn about the park's rich cultural and natural heritage.

niahnationalpark.my

ASIA — BORNEO (MALAYSIA)

80 LAMBIR HILLS NATIONAL PARK

TO VISIT BEFORE YOU DIE BECAUSE

More than 1,000 species of trees can be seen from the vast trail network of this park nestled in the lush heart of Borneo.

Lambir Hills National Park, in the Malaysian state of Sarawak, is a natural treasure trove teeming with biodiversity and breathtaking landscapes. The Lambir Hills are a sanctuary for numerous endangered Malaysian species, such as the Bornean orangutan, the cloud leopard, the Malaysian sun bear, and the critically endangered Bornean slow loris. There are also upwards of 200 bird species soaring through the canopy of this rainforest, ranging from the rhinoceros hornbill to the blue-banded pitta. Lambir Hills is renowned for its network of well-maintained trails, offering visitors the opportunity to explore its varied ecosystems – from leisurely strolls to challenging hikes, there's a trail for every level of adventurer. One of the park's highlights is the hike to the summit of Bukit Lambir, where panoramic views of the surrounding rainforest canopy await those who conquer its ascent.

ASIA BORNEO (MALAYSIA)

81 GUNUNG MULU NATIONAL PARK

TO VISIT BEFORE YOU DIE BECAUSE

High levels of endemism make the park one of the most scientifically studied places on Earth.

Gunung Mulu National Park, found deep within the rainforests of Malaysian Borneo, is a Unesco World Heritage Site renowned for its extraordinary biodiversity – in fact, it's the most studied tropical karst area in the world. It has 17 vegetation zones, including peat swamps, moss forests, riparian forests, and beyond. There are more than 3,500 recognized vascular plant species and an additional 1,500 flowering plant varieties. And then there's the fauna, much of which is endemic to the island. The park holds at least 81 mammals (including 28 species of bats), 270 birds, 55 reptiles, 76 amphibians, 48 fish, 281 butterflies, 458 species of ants, and more than 20,000 invertebrates, though there are likely more of each that just haven't been identified, yet. The park also claims some remarkable caves, including some caverns that are amongst the largest and most intricate in the world. The Sarawak Chamber, for instance, is considered the largest cave chamber by area, capable of fitting several jumbo jets inside.

mulupark.com

ASIA   CAMBODIA

82 PREAH MONIVONG BOKOR NATIONAL PARK

TO VISIT BEFORE YOU DIE BECAUSE

You can take in fabulous mountainous and coastline views from Bokor Hill, in southern Cambodia's Kampot Province.

Those who hike or drive to the summit of Bokor Hill Station (at 3,438 feet above sea level) are rewarded with breathtaking panoramic views of the surrounding Preah Monivong Bokor National Park and the Gulf of Thailand. They will also find a clutch of abandoned buildings, including a church, an old hotel, and crumbling villas. Built in the early 20th century by French colonial settlers as a retreat from the sweltering lowlands for their troops, the dwellings of the ghost town are slowly being reclaimed by nature. It's not possible to get into many of the buildings anymore (with the exception of the church), but the area is interesting to walk around. There are oodles of winding trails (catering to all levels of experience) within the park, making it a destination for hiking, biking, and birdwatching. Along the way, visitors may encounter rare orchids, towering hardwood trees, and elusive wildlife such as macaques, gibbons, leopards, or Asian elephants.

ASIA   CHINA

83   ZHANGYE DANXIA NATIONAL GEOPARK

TO VISIT BEFORE YOU DIE BECAUSE

Famed for its rainbow-hued hills, this park in northwestern China offers one of the world's most eye-catching panoramas.

Known for its red, orange, yellow, pink, and green rock formations, this Unesco Global Geopark is found in the northern foothills of the Qilian Mountains in the Gansu Province of China. The multicolored, striped appearance of the landscape is due to different types of sandstone and minerals (like iron oxide and manganese) being layered on top of each other over millions of years. As the tectonic plates have shifted, rising and falling, the colorful layers have tilted and can now be admired across the hills, particularly in the Linze Danxia Scenic Area. Erosion from wind and rain has further shaped the area, making ravines and pillars. There are various viewing platforms throughout the park, including the "Painted Wall," the "Colorful Clouds," and the "Danxia Landform Pillars." And while the landscape is undoubtedly the star of the show, animal lovers can also expect to be able to see a handful of creatures, including mountain goats, pikas, and golden eagles.

national-parks.org/china/zhangye

ASIA CHINA

84 BADALING NATIONAL FOREST PARK

TO VISIT
BEFORE YOU DIE
BECAUSE

The famous Great Wall of China runs through this park.

Just northwest of Beijing, Badaling National Forest Park boasts natural beauty, historical significance, and cultural richness. At the heart of Badaling National Forest Park stands the Great Wall of China, a testament to human ingenuity and perseverance. Stretching across rugged mountain terrain, the Great Wall snakes its way through the landscape, offering panoramic vistas of rolling hills and verdant valleys. Constructed initially over two millennia ago, the Great Wall long served as a formidable defensive fortification and a symbol of China's enduring resilience against external threats. Travelers can hike on or along the wall, explore ancient watchtowers, visit museums showcasing archaeological artifacts, or attend one of the demonstrations and performances to get a better idea of how the Great Wall has influenced (and continues to influence) the culture of China.

english.visitbeijing.com.cn

ASIA — CHINA

85 GUILIN AND LIJIANG RIVER NATIONAL PARK

TO VISIT BEFORE YOU DIE BECAUSE

A fairytale landscape of sugarloaf mountains, meandering rivers, and rice terraces, it's arguably one of the most famous sights in all of China.

Chances are, if you tell someone in China that you're going to Guilin, they will let you know it's the most beautiful place in the country, even if they have never been. They know because it's the landscape depicted on the back of the 20 yuan bill. Located in the Guangxi Zhuang Autonomous Region of southern China, Guilin is a city that boasts a breathtaking karst topography. Its tall, thin limestone peaks jut dramatically out of the earth, creating a picturesque setting that has long inspired poets and artists. The Li River is the city's lifeblood, and one of the most popular ways to experience Guilin and the surrounding countryside is by going on a day cruise. Along the way, you'll likely see fishermen on traditional rafts and modern boats. You'll also pass the many ancient towns and villages that dot the riverbanks, offering a glimpse into traditional Chinese life.

en.gxzf.gov.cn/regions.html

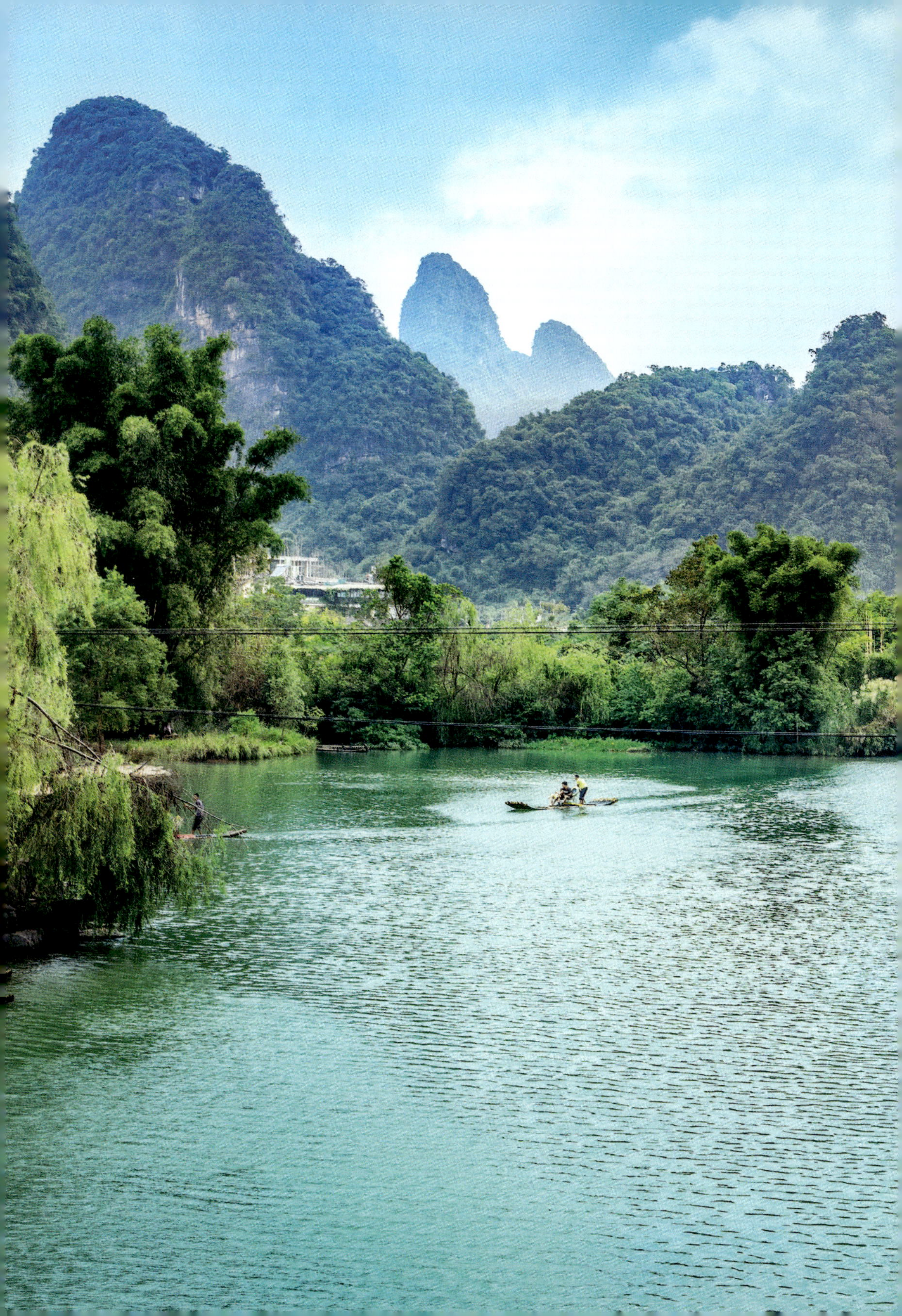

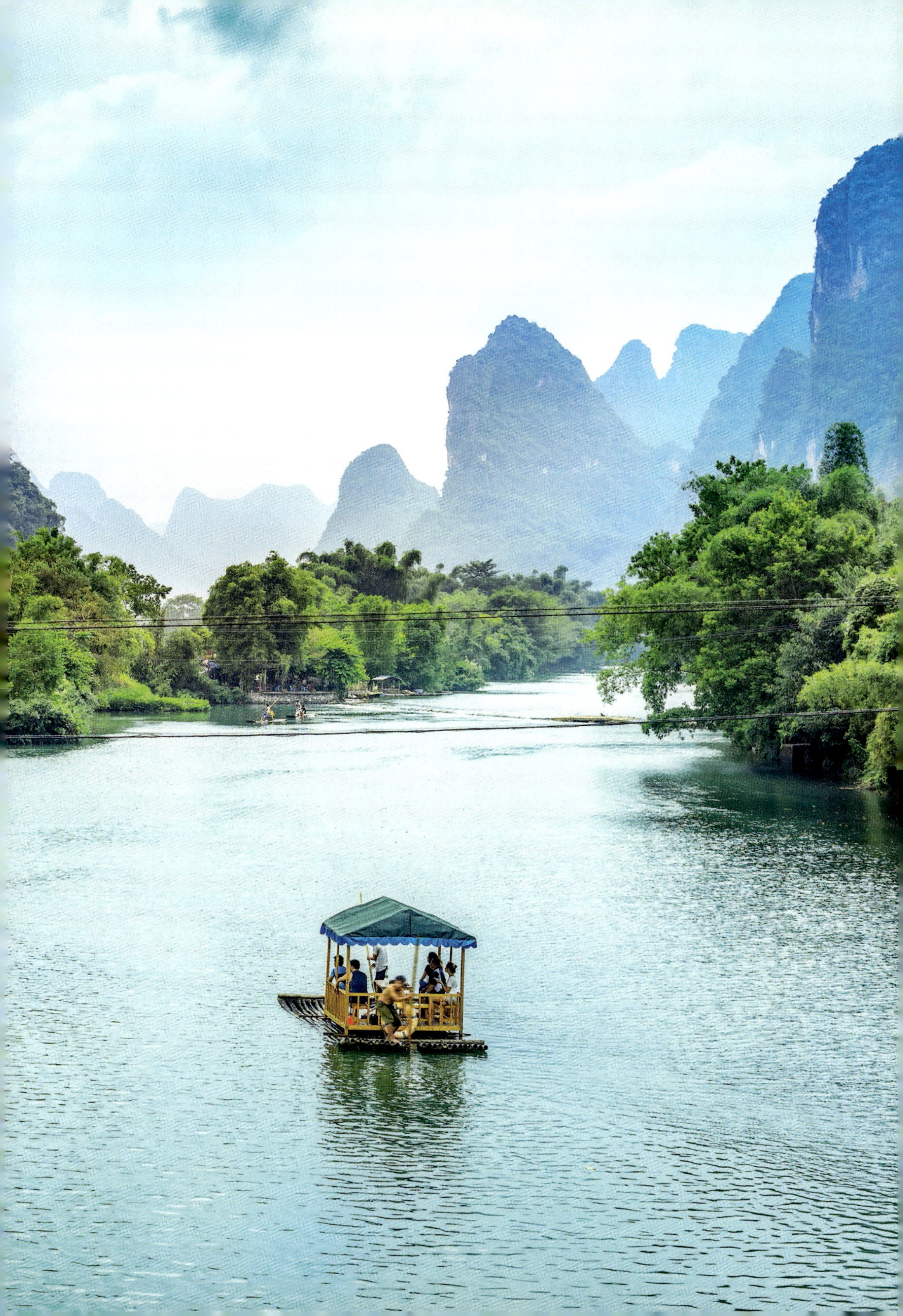

ASIA   ISRAEL

86 MASADA NATIONAL PARK

TO VISIT BEFORE YOU DIE BECAUSE

At its heart is the ancient fortress of Masada, a Unesco World Heritage Site.

Masada National Park, situated on a dramatic plateau overlooking the Dead Sea in Israel, is a place steeped in history and natural beauty. Its most prominent feature is the ancient fortress of Masada, a Unesco World Heritage Site and one of Israel's most interesting landmarks. The fortress of Masada holds immense historical significance, serving as the site of one of the most dramatic events in Jewish history. In the 1st century CE, it was the last stronghold of Jewish rebels who defied Roman rule during the First Jewish-Roman War. Today, Masada National Park welcomes visitors from around the world to explore its archaeological ruins and breathtaking vistas. The park offers various hiking trails, allowing visitors to ascend to the fortress either by foot or via a cable car for a panoramic view of the surrounding Judean Desert landscape and the Dead Sea. Along the trails, hikers encounter remnants of ancient walls, palaces, and storehouses, offering glimpses into the fortress's storied past.

en.parks.org.il/reserve-park/masada-national-park

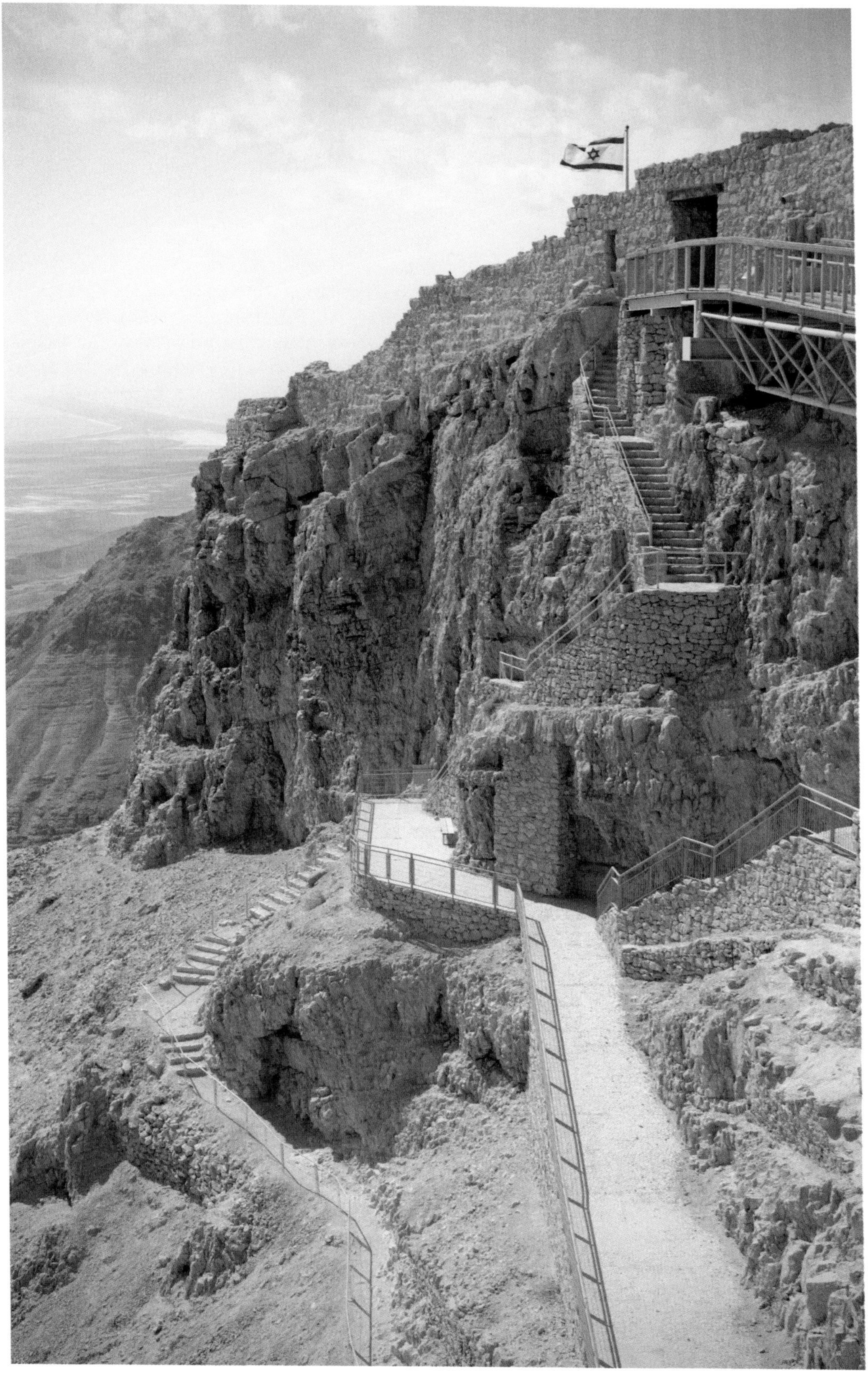

ASIA   INDIA

87   SUNDARBANS NATIONAL PARK

TO VISIT BEFORE YOU DIE BECAUSE

This swathe of mangrove forests in the Ganges delta is the best place to see Bengal tigers in their natural habitat.

Sundarbans National Park, a Unesco World Heritage Site, is a mesmerizing expanse of mangrove forests stretching across the delta of the Ganges, Brahmaputra, and Meghna rivers in India and Bangladesh. In fact, it is the largest tidal halophytic mangrove forest in the world and forms a dense labyrinth of roots and branches that serve as a critical habitat for myriad species, including the estuarine crocodile and the endangered Irrawaddy dolphin. Visitors can navigate the narrow channels and creeks by boat in search of the most famous resident of the national parkland: the Bengal tiger. This elusive predator, perfectly adapted to its watery habitat, roams the mangrove forests in search of prey, its striped coat blending with the dappled sunlight filtered through the dense canopy above. While the likelihood of seeing the big cat is rare, if you are one of the lucky ones, it will be something you talk about for the rest of your life.

ASIA INDIA

88 RANTHAMBORE NATIONAL PARK

TO VISIT BEFORE YOU DIE BECAUSE

Once a private game reserve of the maharajas of Jaipur, this national park in Rajasthan's mountains is home to the elusive Bengal tiger.

Ranthambore National Park is a jewel nestled in the Aravalis and Vindhya mountains of Rajasthan. Spread across an area of approximately 244 square kilometers, it stands as one of the largest and most renowned national parks in Northern India. The park's flora is characterized by dry deciduous forests dominated by species such as dhok, mango, banyan, and pipal trees. These dense woodlands provide shelter and sustenance to a plethora of wildlife, including the park's star attraction: its population of Bengal tigers. With an estimated 70 to 80 individuals, these apex predators roam freely within the park's boundaries, offering visitors a rare opportunity to witness these magnificent creatures in their natural habitat. The park's tiger sightings are facilitated by its network of well-maintained safari tracks and strategically located water bodies, where tigers often congregate, especially during the scorching summer months. Other animals include the Indian leopard, striped hyena, sloth bear, wild boar, sambar deer, and chital.

www.tourism.rajasthan.gov.in/sawaimadhopur.html#ranthambore

ASIA　　　　　　　　JAPAN

89 FUJI-HAKONE-IZU NATIONAL PARK

TO VISIT
BEFORE YOU DIE
BECAUSE

The highlights of Japan's most visited park include rugged coastlines, hot springs, a lush peninsula and, of course, the mighty Mount Fuji.

Mount Fuji, Japan's highest peak, is revered by locals for its majestic beauty and cultural significance. Towering at 12,388 feet, the perfectly symmetrical cone-shaped mountain is the symbol of Japan. It's only possible to summit the snow-covered mountain for two months in the summer, yet each season in recent years, over 400,000 people have reached the top. Beyond Mount Fuji, the sprawling park also consists of Fuji Five Lakes, Hakone, the Izu Peninsula, and the Izu Islands. The Hakone region is famous for its hot springs, stunning views of the surrounding mountains and stunning lakes – the nearby Lake Ashi, a scenic crater lake formed by volcanic activity, for instance, offers visitors the chance to take leisurely boat cruises while admiring the reflection of Mount Fuji on its tranquil waters. Then there's the Izu Peninsula, which boasts rugged coastlines, sandy beaches, and lush subtropical forests. Throughout the park, visitors can immerse themselves in Japan's rich cultural heritage by visiting traditional villages, ancient shrines, and historic sites.

www.japan.travel/national-parks/parks/fuji-hakone-izu

ASIA / JORDAN

90 WADI RUM PROTECTED AREA

TO VISIT BEFORE YOU DIE BECAUSE

Bedouin traditions, sand dunes, huge rocky outcrops, and desert vistas make this a truly magical destination.

Wadi Rum Protected Area, also known as the "Valley of the Moon," is a vast desert wilderness covering an area of over 450 square miles. This remarkable landscape of towering sandstone mountains, sweeping dunes, and ancient valleys is one of the most breathtaking and mesmerizing destinations in the Middle East. From the imposing Jabal Rum, the highest peak in Jordan, to the towering natural arches that dot the terrain, every vista in Wadi Rum is a testament to the power of nature's sculpting hand. Visitors to Wadi Rum can explore the park through a variety of activities, including jeep tours, camel rides, hiking, rock climbing, and camping under the stars. There are also opportunities to take a guided tour led by a local Bedouin guide, who can offer unique insight into the area's history and culture, in addition to the natural wonders.

wadirum.jo

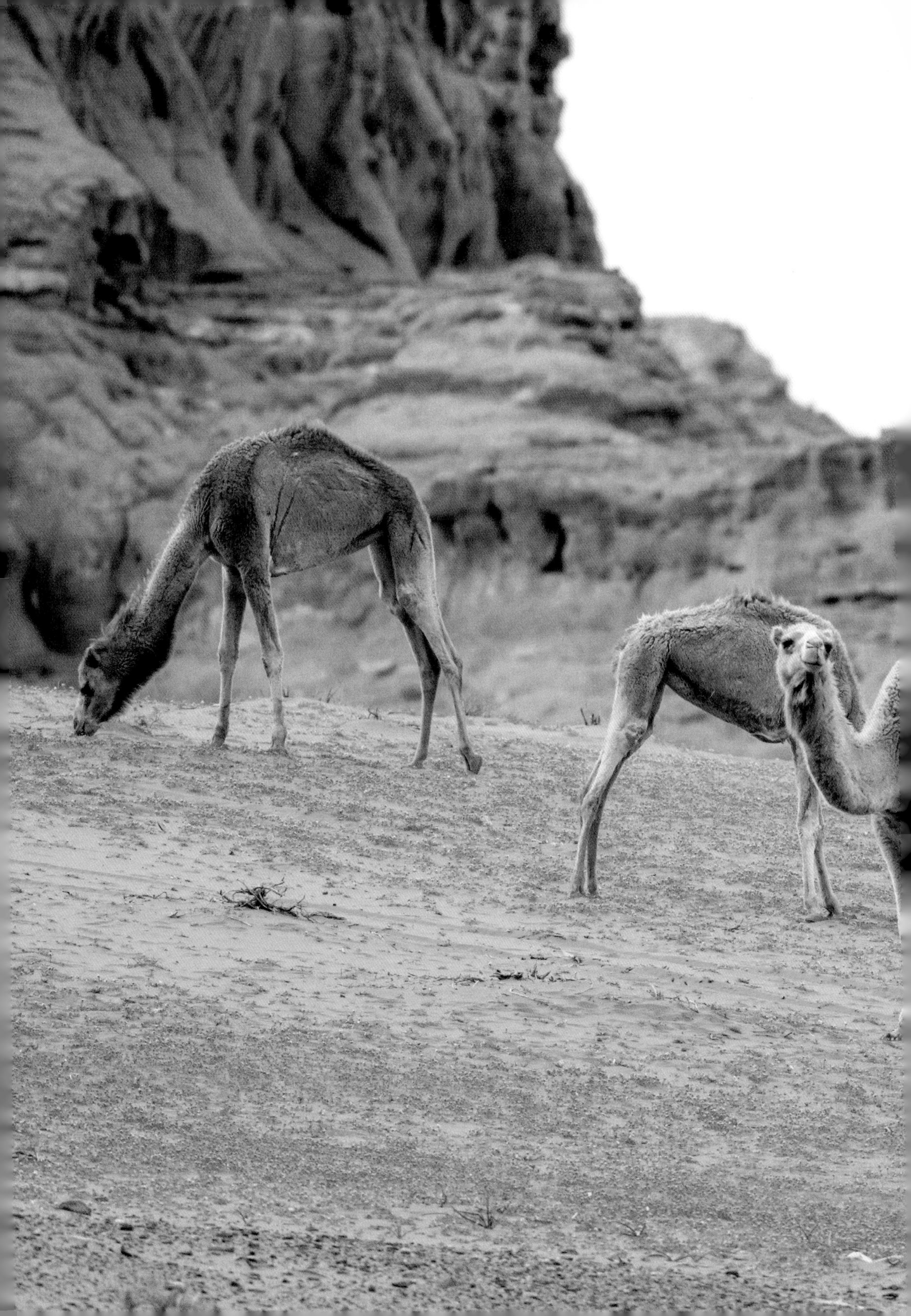

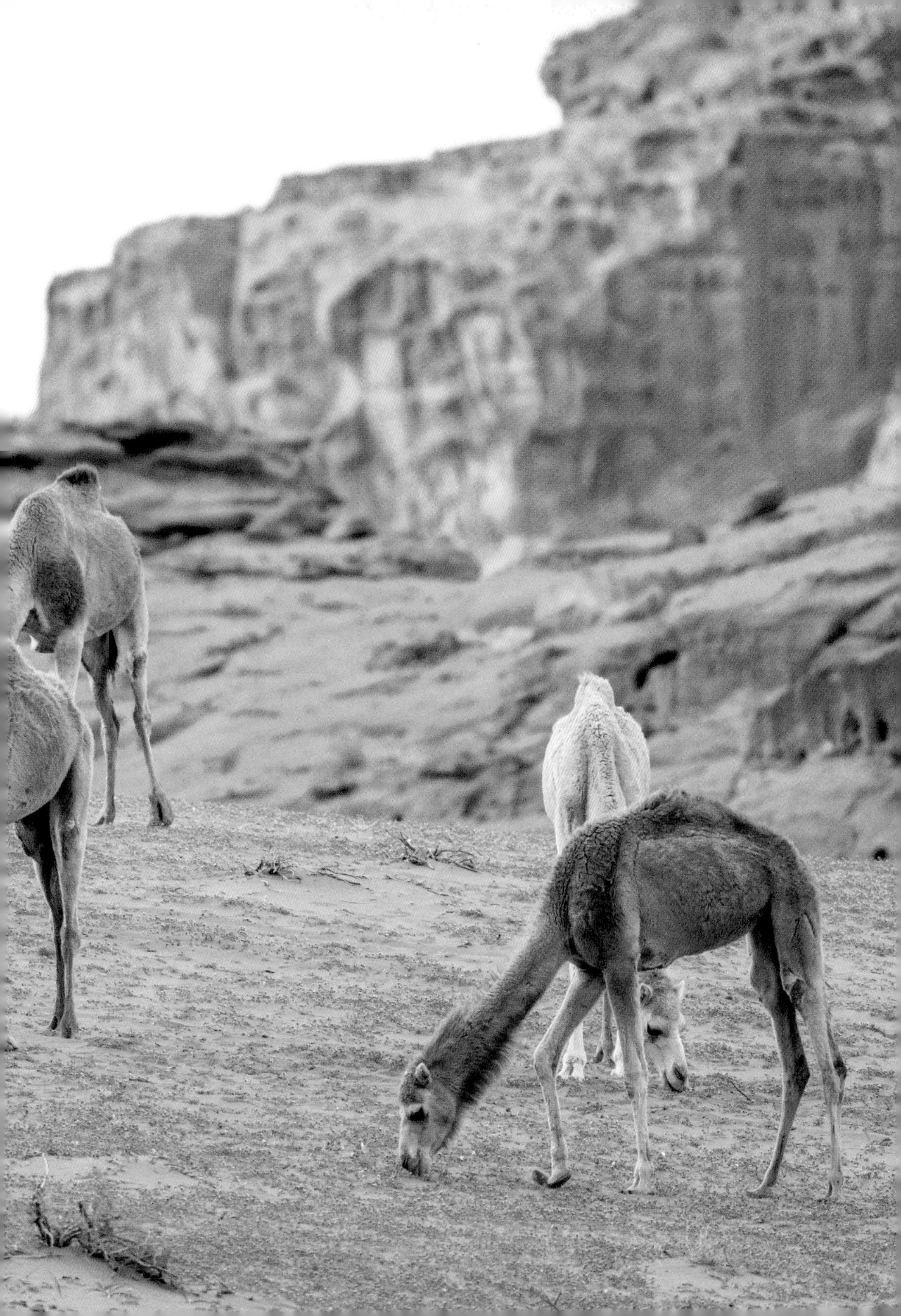

ASIA — INDONESIA

91 KOMODO NATIONAL PARK

TO VISIT BEFORE YOU DIE BECAUSE

The world's largest reptile, the Komodo dragon, rules this string of islands in the center of the Indonesian archipelago.

A living relic from prehistoric times and the world's largest lizard species, Komodo dragons are the big draw of this Indonesian national park. These formidable creatures, known for their immense size (reaching approximately 10 feet in length and weighing more than 150 pounds) and predatory prowess (they can see objects from nearly 1,000 feet away and can run and swim at speeds of 12 miles per hour), roam freely across the rugged terrain. Situated within the Lesser Sunda Islands, this national park encompasses several islands, with the three largest being Komodo, Rinca, and Padar. It also boasts a diverse range of ecosystems, including pristine beaches, lush forests, and vibrant coral reefs. Snorkeling and diving enthusiasts from around the world are drawn to the park due to its rich marine environment, with crystal-clear waters where they can explore an underwater wonderland teeming with colorful marine life, including manta rays, sea turtles, and countless species of fish.

www.indonesia.travel/gb/en/destinations/bali-nusa-tenggara/labuan-bajo/komodo-national-park.html

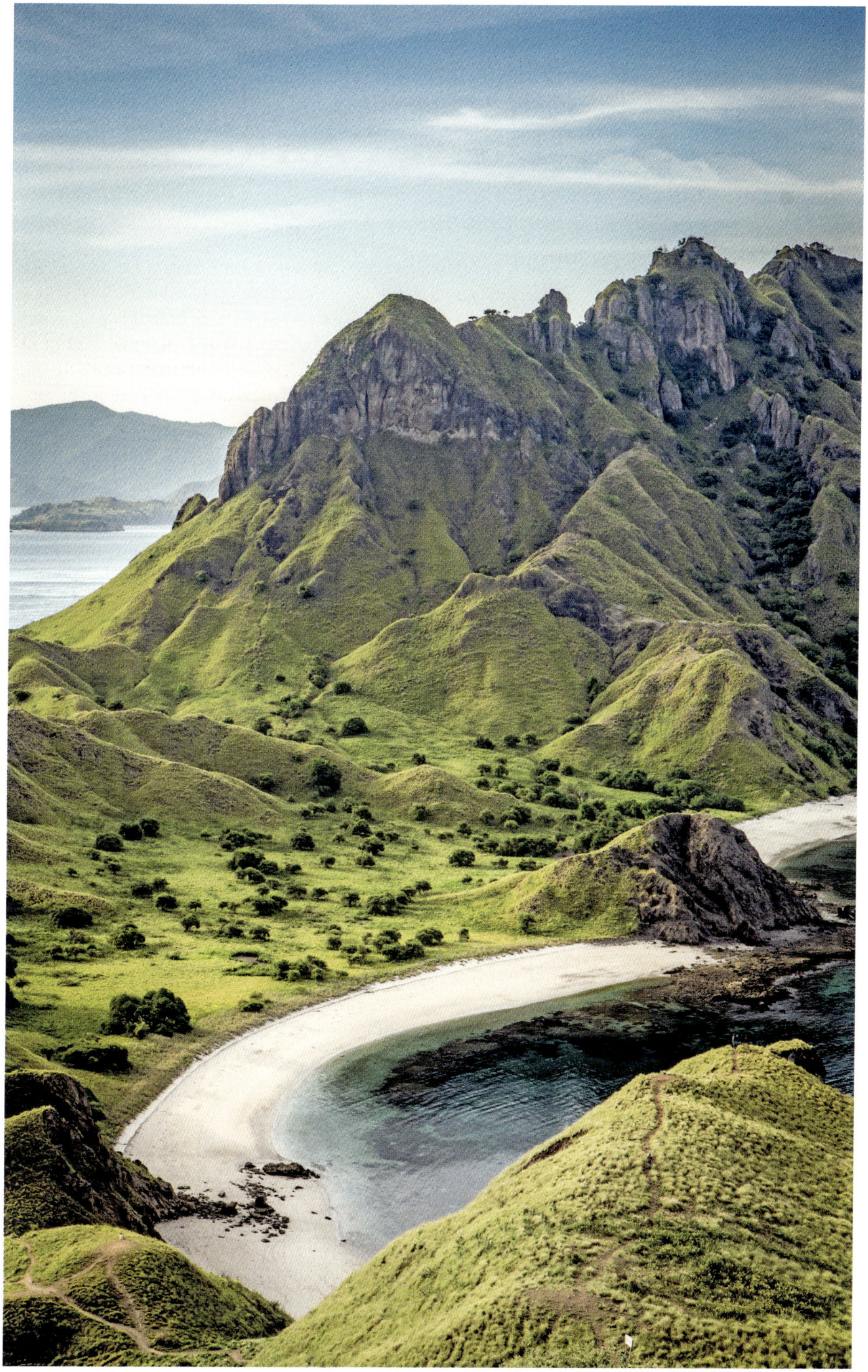

ASIA — KAZAKHSTAN

92 ALTYN-EMEL NATIONAL PARK

TO VISIT BEFORE YOU DIE BECAUSE

Wild horses roam before a backdrop of colorful mountains and sand dunes.

If you like unusual landscapes and underexplored destinations, Altyn-Emel National Park is for you. One of the park's most iconic features is the Singing Sand Dune, a massive golden-hued mountain of sand. When the wind sweeps across its surface, it produces a haunting melody that reverberates through the desert. Among the other famous landmarks is the Aktau Mountains, a surreal landscape of vividly colored rock formations, where fossils of plants and animals from more than 25 million years ago have been found. Altyn-Emel is also home to a wide variety of animals adapted to survive in this arid environment, including the desert-eared hedgehog, the steppe eagle, the Siberian ibex, the Corsac fox, the black stork, and the Tien Shan brown bear. Adventurous travelers can look for wildlife and behold the natural beauty of the rugged park on a guided jeep safari or on horseback.

kazakhstan.travel/en/tourist-spot/31/altyn-emel-national-park

ASIA   MONGOLIA

93 GORKHI-TERELJ NATIONAL PARK

TO VISIT BEFORE YOU DIE BECAUSE

This stunning landscape, known as the Switzerland of Mongolia, is full of surprises, from granite formations to crystal-clear streams.

Just northeast of Ulaanbaatar, Mongolia's capital city, Gorkhi-Terelj National Park is known for its verdant forests of Siberian larch and birch, meandering rivers and streams, and high granite mountains, all of which provide a haven for an array of wildlife including deer, foxes, and even the elusive Siberian ibex. One of the park's most iconic landmarks is the striking Turtle Rock, a massive granite formation shaped like a turtle emerging from the earth. Locals believe it is a good luck symbol. Visiting it, hiking through the green meadows, and going for long horseback rides are all reason enough to visit. However, for travelers yearning for a taste of traditional Mongolian life, Gorkhi-Terelj offers unique opportunities to stay in traditional ger camps and experience the nomadic lifestyle firsthand. Here, visitors can savor hearty meals of traditional Mongolian cuisine and even participate in activities such as milking yaks or learning the art of archery from local experts.

ASIA NEPAL

94 SAGARMATHA NATIONAL PARK

TO VISIT BEFORE YOU DIE BECAUSE

Sagarmatha is the Nepalese name for Mount Everest, the world's tallest mountain, which is best admired from this High Himalayan paradise.

Few mountains captivate the imagination quite like Everest. At 29,032 feet, it's the tallest peak in the world. It attracts mountaineers and trekkers from around the globe, drawn by the challenge of conquering the summit. However, even just the journey to Everest Base Camp takes trekkers through stunning landscapes and Sherpa villages, including Namche Bazaar, Tengboche, and Khumjung, offering a glimpse into the unique way of life of the Sherpa people, known for their resilience, hospitality, and mountaineering prowess. The hike also offers unparalleled views of the world's highest peaks and a lifetime of bragging rights. Even the lowest elevations of the park are incredible, adorned with dense forests of pine, birch, and rhododendrons. It is this area of the park that is a haven for rare species such as the elusive snow leopard, red panda, Himalayan tahr, and musk deer, which freely roam its remote valleys.

www.snp.gov.np

95 CHITWAN NATIONAL PARK

TO VISIT BEFORE YOU DIE BECAUSE

Set in the country's subtropical lowlands, this World Heritage Site is a lush jungle home to tigers, rhinos, king cobras, and crocodiles.

When you hear the word "Nepal," what comes to mind? For many, it is the ragged Himalaya mountain range, where the famed Everest and Annapurna trails beckon to hikers and climbers. However, Chitwan is often referred to as "the other Nepal." Here in the subtropical lowlands of south-central Nepal, the landscape is more meadows and marshlands, providing a haven for diverse flora and fauna. Chitwan translates to "Heart of the Jungle," and within its boundaries, endangered species thrive. Among its inhabitants are leopards, gharial crocodiles, one-horned rhinoceros, Asian elephants, sloth bears, and Royal Bengal tigers, the largest big cat in the world. The park also shelters over 500 species of birds, ranging from kingfishers with their vibrant plumage to hornbills identifiable by their long down-turned beak. In Chitwan National Park most of the activities center around wildlife and bird spotting – visitors can sign up for jeep or ox cart safaris, dug-out canoe rides, or a walking safari if they're really brave.

ASIA PHILIPPINES

96 HUNDRED ISLANDS NATIONAL PARK

TO VISIT BEFORE YOU DIE BECAUSE

The country's first national park, this protected area in Pangasinan's Lingayen Gulf offers the ultimate island-hopping experience.

Don't let the name fool you – there are way more than a hundred islands at Hundred Islands National Park, scattered across the gin-bottle blue waters of the Lingayen Gulf in the Philippines. There are actually 124 at low tide and 123 islands at high tide. However, of these scenic isles, only four have been developed for tourism since the park was established in 1940: Children's Island, Governor Island, Marcos Island, and Quezon Island. Each island within the park boasts its own unique features, ranging from lush greenery and towering limestone cliffs to pristine white sandy beaches and hidden coves. Exploring the park offers visitors a wealth of activities, including island-hopping, swimming, trekking, or snorkeling to see the tropical fish and colorful coral reefs just off the shore. Beyond sea life, the most common animals in the area are birds, including the Philippines duck, white-eared brown-dove, Philippine coucal, elegant tit, lemon-throated leaf-warbler, and grey-backed tailorbird.

philippines.travel/activities/go-swimming-and-island-hopping-hundred-islands

ASIA RUSSIA

97 STOLBY NATURE RESERVE

TO VISIT BEFORE YOU DIE BECAUSE

Also known as Krasnoyarsk Pillars, Russia's most visited national park attracts climbers and hikers thanks to its dramatic rock formations.

Just because this nature reserve is located just across the river from Krasnoyarsk's city center and is the most visited national park in Russia, it doesn't mean that this protected land isn't wild and full of adventure. The park, which was established in 1945, is named for its most striking feature: a collection of rock formations sculpted over eons by the relentless forces of nature, that punctuate the landscape. Rock climbing enthusiasts flock to Stolby to test their skills on the challenging granite cliffs that define the landscape. There are over 700 vascular plants and upwards of 250 kinds of mosses found within the park. There are also nearly 300 species of mammals, including sable, Siberian musk deer, hazel grouse, steppe polecat, long-tailed ground squirrel, and some impressive birds of prey, such as the osprey, golden eagle, saker falcon, peregrine falcon, and others.

whc.unesco.org/en/tentativelists/5113

ASIA SOUTH KOREA

98 SEORAKSAN NATIONAL PARK

TO VISIT BEFORE YOU DIE BECAUSE

This park is home to the oldest Zen Buddhist temple on the planet, as well as serene valleys and rugged granite peaks.

South Korea's third largest national park offers myriad reasons to visit, though chief among them is seeing the park's centerpiece, Mount Seoraksan. One of the tallest peaks in South Korea at 5,604 feet above sea level, it has the distinction of becoming Korea's first Unesco Biosphere Reserve. Its rugged granite peaks, lush forests, cascading waterfalls, and serene valleys create an awe-inspiring panorama that changes dramatically with the seasons. In spring, vibrant cherry blossoms blanket the mountainsides, while in autumn, fiery hues of red and gold paint the landscape. Traversing the park's network of hiking trails is a captivating journey, offering breathtaking vistas at every turn. One of the most popular routes is the Ulsanbawi Rock Trail, which leads to towering rock formations with panoramic views of the surrounding valleys and coastline. For the more adventurous, the challenging Daecheongbong Peak trail provides a thrilling ascent to the park's highest point.

ASIA SOUTH KOREA

99 NAEJANGSAN NATIONAL PARK

TO VISIT BEFORE YOU DIE BECAUSE

The smallest park in South Korea packs the biggest autumnal punch and offers access to stunning waterfalls and a thousand-year-old Budhhist sanctuary.

Renowned for its breathtaking autumnal foliage, Naejangsan National Park is a magnet for visitors who flock to witness the forest transform into a vibrant canvas of red, orange, and gold. As the fall season sets in, the park undergoes a stunning metamorphosis, with the leaves of maple trees, oak trees, and other deciduous species painting the landscape in a kaleidoscope of colors. Apart from its autumnal splendor, Naejangsan National Park, which was established in 1971 and remains the smallest of South Korea's parks, offers a wealth of outdoor activities and natural attractions to explore throughout the year. Among the park's notable landmarks is Naejangsan, a stately mountain that rises to an elevation of 2,503 feet and is crowned by the historic Naejangsa Temple, a revered Buddhist sanctuary that dates back over a thousand years. The park also boasts a handful of beautiful waterfalls, including the famed Dodeokpokpo Falls and Geumseonpokpo Falls.

national-parks.org/south-korea/naejangsan

ASIA SOUTH KOREA

100 JIRISAN NATIONAL PARK

TO VISIT
BEFORE YOU DIE
BECAUSE

The largest national park in South Korea, this natural and spiritual landscape is peppered with sacred sites and ancient temples.

Locally, the name Jirisan translates to "the mountain of the odd and wise people," and it's that mountain, the second tallest in the country (and the tallest on the mainland, at 6,283 feet), that is the nucleus of the park. Just under 25 miles of trails spider through the park (including on Jirisan), offering plenty of ways to work up a sweat and encounter wildlife like elk, roe deer, wildcat, and Asiatic black bear. Jirisan National Park is not only a haven for nature enthusiasts but also a spiritual retreat deeply ingrained in Korean culture and history. Scattered throughout the park are ancient temples, hermitages, and sacred sites, including Hwaeomsa Temple, one of the most significant Buddhist temples in Korea, dating back roughly 1,500 years. These cultural treasures serve as reminders of the park's role as a place of pilgrimage and reflection for generations of Koreans.

ASIA SRI LANKA

101 YALA NATIONAL PARK

TO VISIT BEFORE YOU DIE BECAUSE

Sri Lanka's most famous park is home to the highest concentration of leopards in the world, and a multitude of different habitats.

Yala has it all: dense monsoon forests, grassy plains, scrublands, brackish lagoons, sandy beaches, marine and freshwater wetlands, and rocky outcrops. Together these habitats provide homes for a wide range of animals, including elephants, sloth bears, water buffalos, saltwater crocodile, and at least 215 bird species (six of which are endemic to Sri Lanka), ranging from lesser flamingos to gray hornbills. Though perhaps the most famous inhabitants of this national park are its Sri Lankan leopards. In fact, Yala boasts the highest concentration of leopards in the world – there are currently estimated to be more than 100 in the park, a number that is growing each year. With all that going on, it makes sense that Yala, the second-largest national park in the country, is the most visited. Many visitors to Yala choose to explore the protected land via a Jeep safari, where experienced guides cruise around the park, pointing out the wildlife and giving mini zoology lessons.

www.yalasrilanka.lk

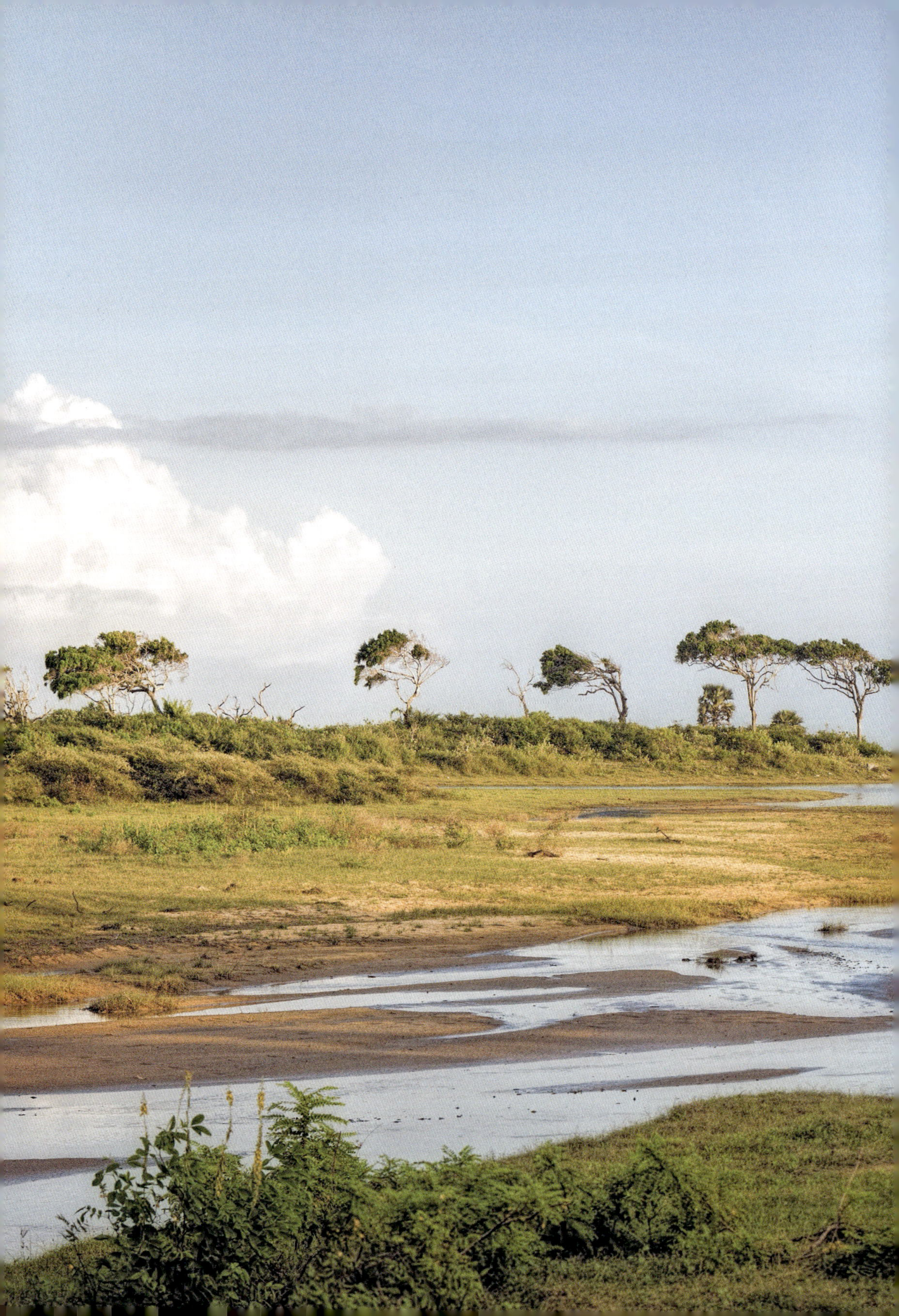

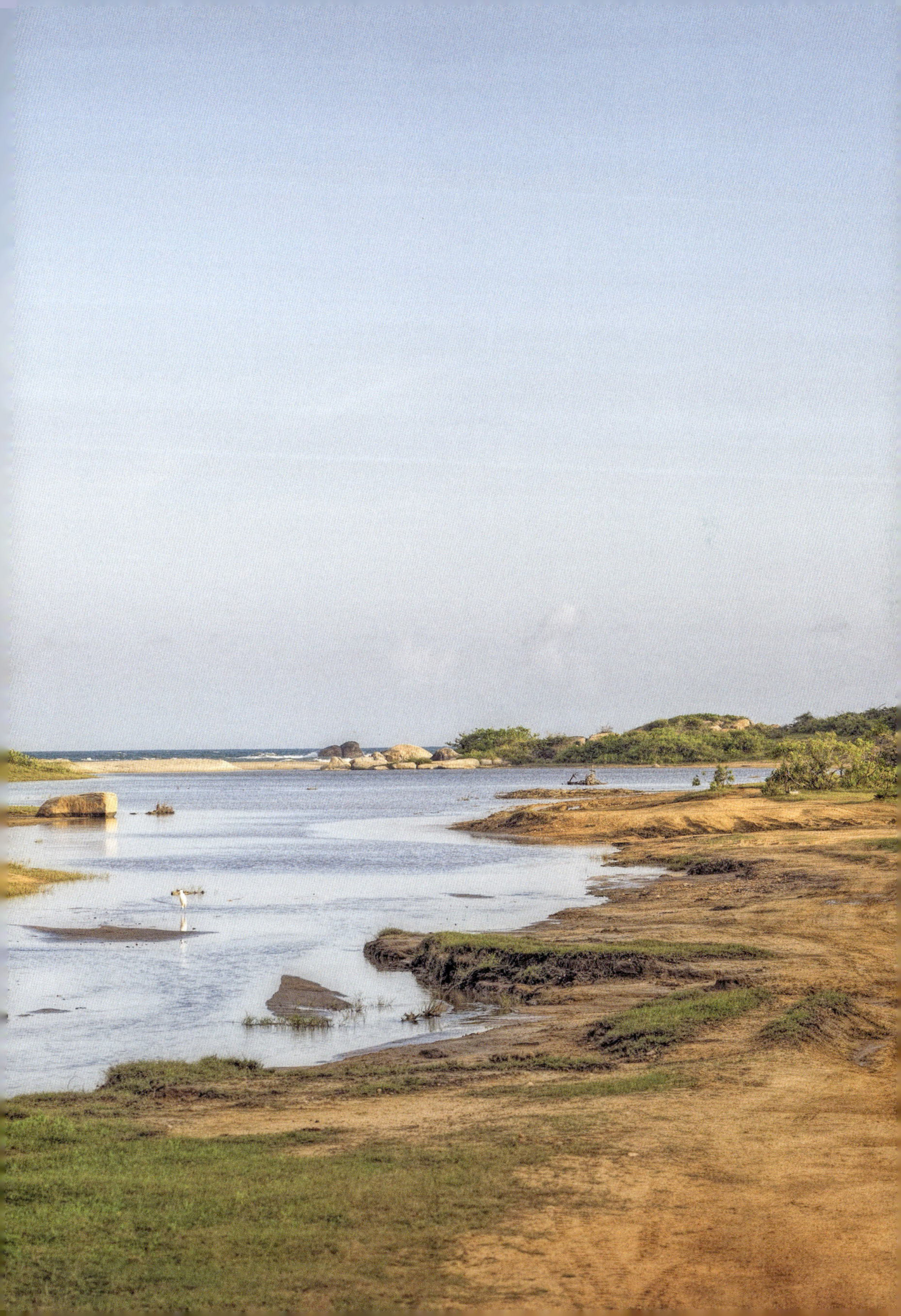

ASIA   TAIWAN

102 TAROKO NATIONAL PARK

TO VISIT BEFORE YOU DIE BECAUSE

To see a stunning marble gorge carved by a fast-moving river, a swallow-filled cave, and picturesque coastal cliffs and waterfalls.

Covering 360 square miles, this is arguably Taiwan's most popular national park. Named for the magnificent Taroko Gorge, a natural masterpiece of sheer marble walls carved over millions of years by the Liwu River, this protected landscape is nothing short of dramatic. Some of the most scenic viewpoints are the Swallow Grotto (where birds number in the thousands), Zhuilu Cliff, and Qingshui Cliffs. One of the park's other landmarks is the Eternal Spring Shrine, a memorial for veterans, which is perched precariously on a picturesque cliff. Additional cultural highlights include the Changuang Temple, Liufang Bridge, and Cimu Bridge (also known as the Motherly Devotion Bridge). Other favorite sites include the various waterfalls, such as the Baiyang Waterfall, which is accessible by a flat and easy hike and visible from a bridge. There is also the Changchun Waterfall, Lushui Waterfall, and Yindai Waterfall.

www.taroko.gov.tw/en

ASIA THAILAND

103 KHAO SOK NATIONAL PARK

TO VISIT BEFORE YOU DIE BECAUSE

Spread over 740 square kilometers in southern Thailand, it stands as one of the oldest rainforests in the world, dating back over 160 million years.

There are few places in Thailand, or Asia for that matter, that are as dreamy as Khao Sok National Park. After the construction of the Ratchaprapha Dam, a vast reservoir now known as Cheow Lan Lake was formed. The emerald waters of the lake are surrounded by towering limestone cliffs, creating an otherworldly landscape that is best explored by a long-tail boat or kayak. Hidden among the lake's islands are floating bungalows, offering visitors the chance to spend the night surrounded by the tranquil beauty of the jungle. It's a popular destination for adventure travelers and nature enthusiasts. The dense jungle is home to a diverse array of wildlife, including elusive species such as Asian elephants, Malayan sun bears, and even the rare clouded leopard. Birdwatchers might also be able to spot the vibrant plumage of hornbills, kingfishers, and other exotic avian species flitting through the canopy.

national-parks.org/thailand/khao-sok

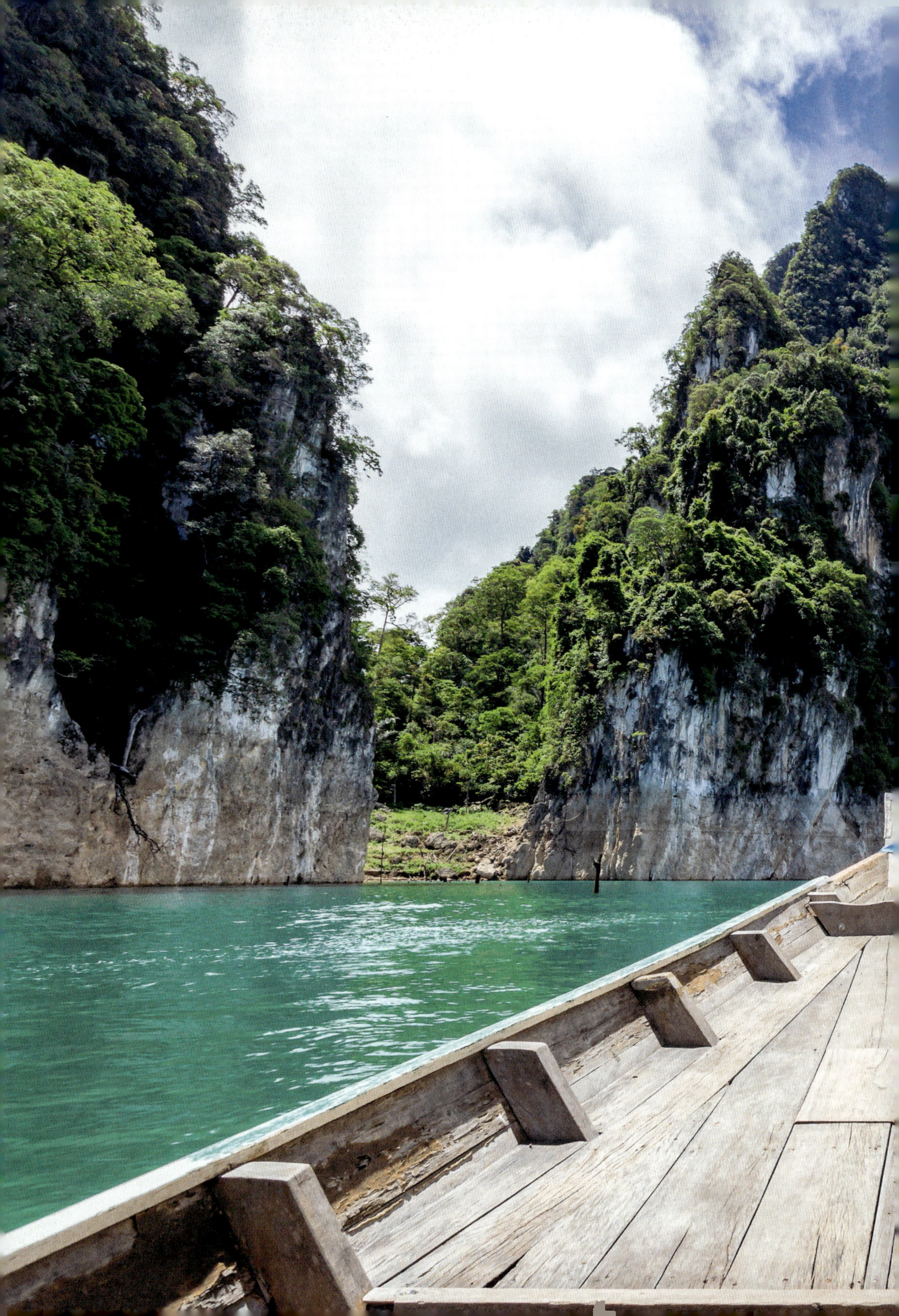

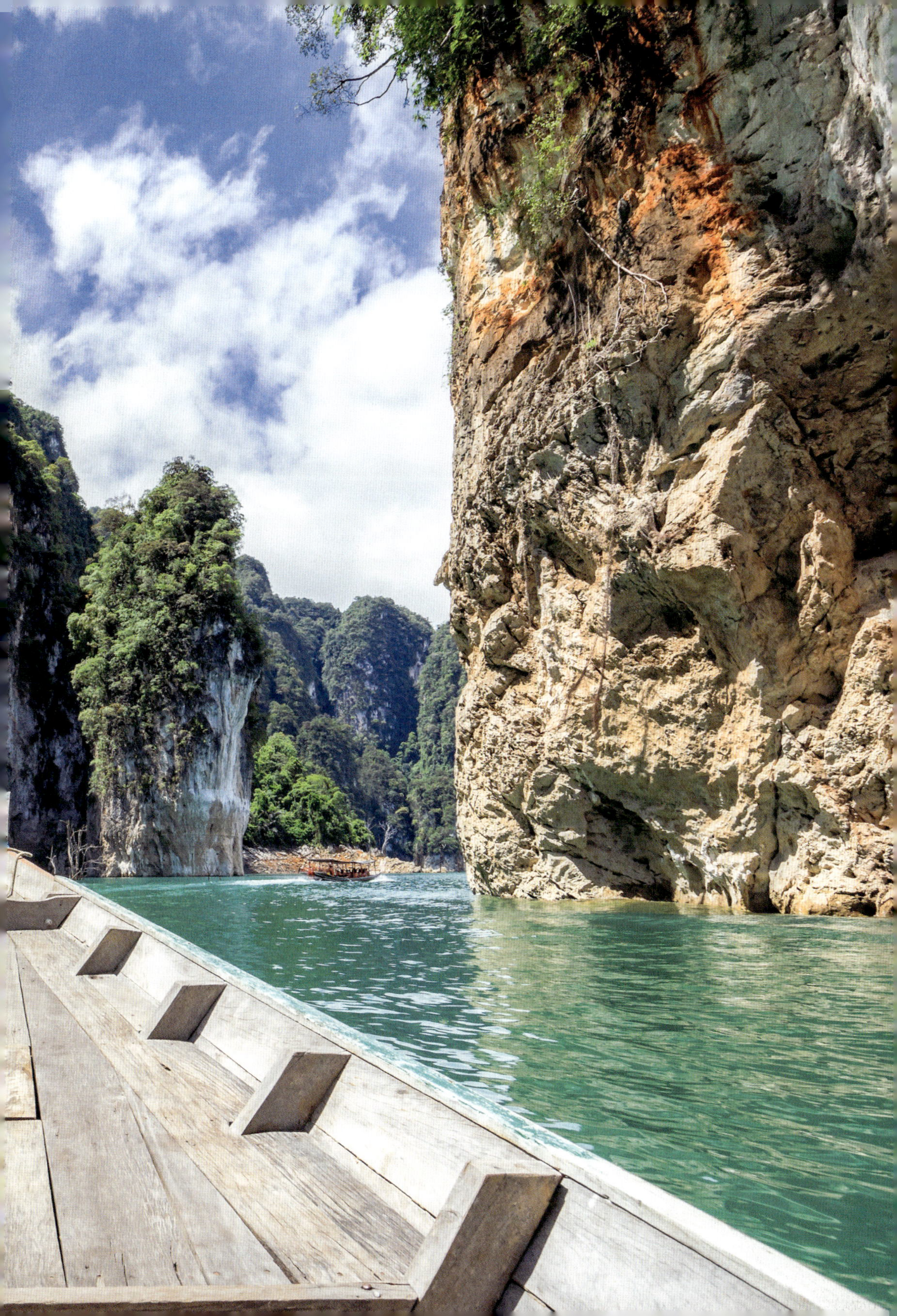

ASIA　　　　　　　　TURKEY

104　GÖREME NATIONAL PARK

TO VISIT BEFORE YOU DIE BECAUSE

Best admired from a hot air balloon, Cappadocia's fantastical, lunar landscape is dotted with fairy chimneys and cave dwellings.

When the weather is right, hundreds of hot air balloons take off from Cappadocia, Turkey, at sunrise and sail over Göreme National Park, offering travelers a bird's-eye view of the landscape. But that's not even the dreamiest or most surreal thing about the Unesco World Heritage Site and protected parkland. The park is characterized by its distinctive rock formations, known as fairy chimneys. Millions of years ago, volcanic eruptions blanketed the area in ash, which solidified into a type of soft rock called "tuff." In the centuries that followed, wind and water erosion shaped the towering pillars, some reaching over 100 feet in height, creating an enchanting lunar-like landscape. Then, during the Roman period, persecuted Christians fled to the area and realized the rock was malleable enough to be hollowed out. Within these rock formations, they carved out spaces for churches, homes, stables, and more. In some places, you'll even find stunning frescoes, giving a glimpse into the religious and artistic traditions of that time.

105 WADI WURAYAH NATIONAL PARK

TO VISIT BEFORE YOU DIE BECAUSE

This dramatic desert landscape offers a unique year-round waterfall and the chance of seeing an animal unseen to human eyes for nearly 30 years.

This protected area is the first national park in the United Arab Emirates, having been established in 2009. It is actually the only terrestrial protected area in the Emirate of Fujairah and it covers more than 20 percent of the emirate's total area. Here visitors will also find wildlife such as the mountain gazelle, caracal, Gordon's wildcat, and Arabian tahr (an endangered goat-like animal). It is possible that the Arabian leopard still roams these parts – while the last sighting of the elusive animal was in 1995, paw prints have been seen in recent years. The United Arab Emirates, only native orchid and only year-round waterfall (it is one of the largest and only freshwater sites in the Middle East) are found here. There are also archaeological sites that date back thousands of years, including ancient petroglyphs and remnants of human settlements.

www.moccae.gov.ae/en/open-data/ecotourism/wadi-wurayah-national-park.aspx

ASIA VIETNAM

106 PHONG NHA-KẺ BÀNG NATIONAL PARK

TO VISIT BEFORE YOU DIE BECAUSE

An adventurer's dream, this park in north-central Vietnam boasts an expansive underground network with spectacular stalactites and stalagmites.

Encompassing more than 60 miles of caves and underground rivers, Phong Nha-Kẻ Bàng National Park, located in the middle of the Annamite Mountain Range in Quang Binh province of Vietnam, is one of the most expansive limestone karst ecosystems in the world (which is part of the reason the park was inscribed as a Unesco World Heritage Site in 2003). Beyond the caves, the park's surface is a mosaic of dense jungles, meandering rivers, and cascading waterfalls. That landscape allows for a wide range of wildlife, including over 150 mammals, 100 reptiles, 50 amphibians, 300 birds, and 170 fish species. Some of the most famous inhabitants of this park include rare species, like the Indochinese tiger and the saola. The saola, which is the cousin of cattle but more closely resembles an antelope, is also sometimes called "Asian unicorn," an appellation it earned due to its rarity and gentle nature.

vietnam.travel/things-to-do/complete-guide-phong-nha

ASIA · VIETNAM

107 CAT BA NATIONAL PARK

TO VISIT BEFORE YOU DIE BECAUSE

On this Halong Bay island is the opportunity to witness the most endangered primate in the world, the golden-headed langur.

Situated on Cat Ba Island, the largest island in the Unesco-listed Halong Bay archipelago, Cat Ba National Park encompasses an array of ecosystems, ranging from lush forests to rugged limestone karsts to vibrant coastal habitats. The diversity of the biospheres means a slew of unique animals are present in the park, including macaques, deer, the giant black squirrel, civets, and the endangered Cat Ba or golden-headed langur, a species of monkey found only on the island. There are also at least 70 species of birds, such as hawks, hornbills, and cuckoos. Excellent hiking is available on the island, with a network of well-maintained trails. One option is the short but steep trek to the summit of Ngu Lam Peak, where the views of the jungle are expansive. Alternatively, it's possible to hike to the village of Viet Hai, where you can spend the night. However, if you're more into ocean adventures, Cat Ba National Park also boasts a stunning coastline dotted with secluded beaches and coves, providing idyllic spots for swimming, snorkeling, and kayaking.

catbanationalpark.vn

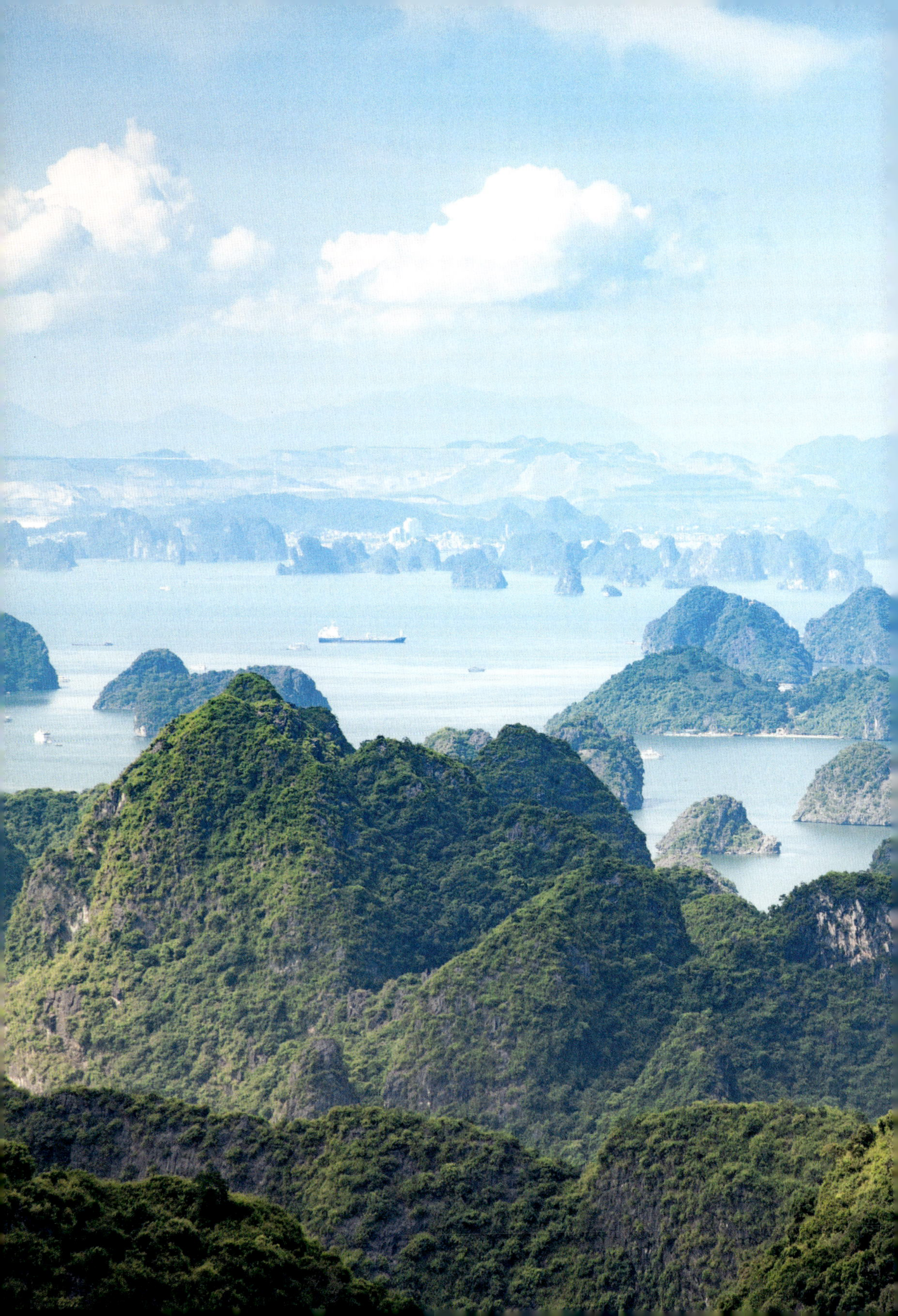

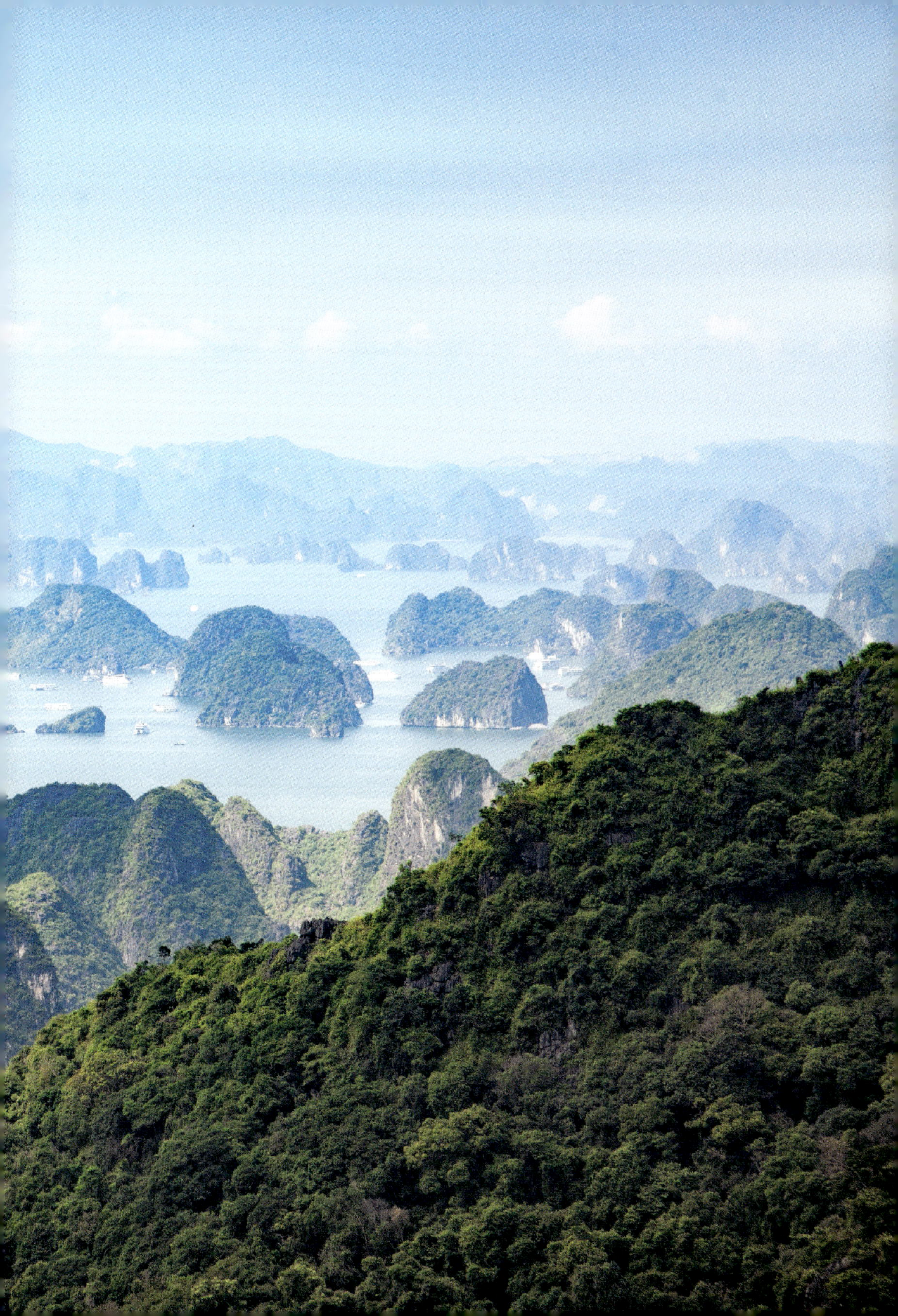

EUROPE ALBANIA

108 BUTRINT NATIONAL PARK

TO VISIT BEFORE YOU DIE BECAUSE

This is a remarkably well-preserved ancient city found in stunning nature, near Sarandë on Albania's turquoise Ionian coast.

Located on the southwestern coast of Albania, this national park and Unesco World Heritage Site offers a chance to step back in time. At its core lies the ancient city of Butrint, whose origins trace back to the 7th century BC. Over millennia, Butrint thrived as a vital hub for civilizations, including the Greeks, Romans, Byzantines, Venetians, and Ottomans. Each left its mark, resulting in a layered tapestry of architectural marvels, including temples, amphitheaters, aqueducts, fortifications, fountains, and two castles. The city's strategic location along the Adriatic coast made it a prized possession and a testament to the ebb and flow of empires throughout history. Numerous shady trails crisscross the park, which is surrounded by wetlands. There's also an excellent Museum of the Ancient City, found within the Acropolis castle, which shares a chronological history of Butrint, starting in the Iron Age and progressing through the Late Middle Ages.

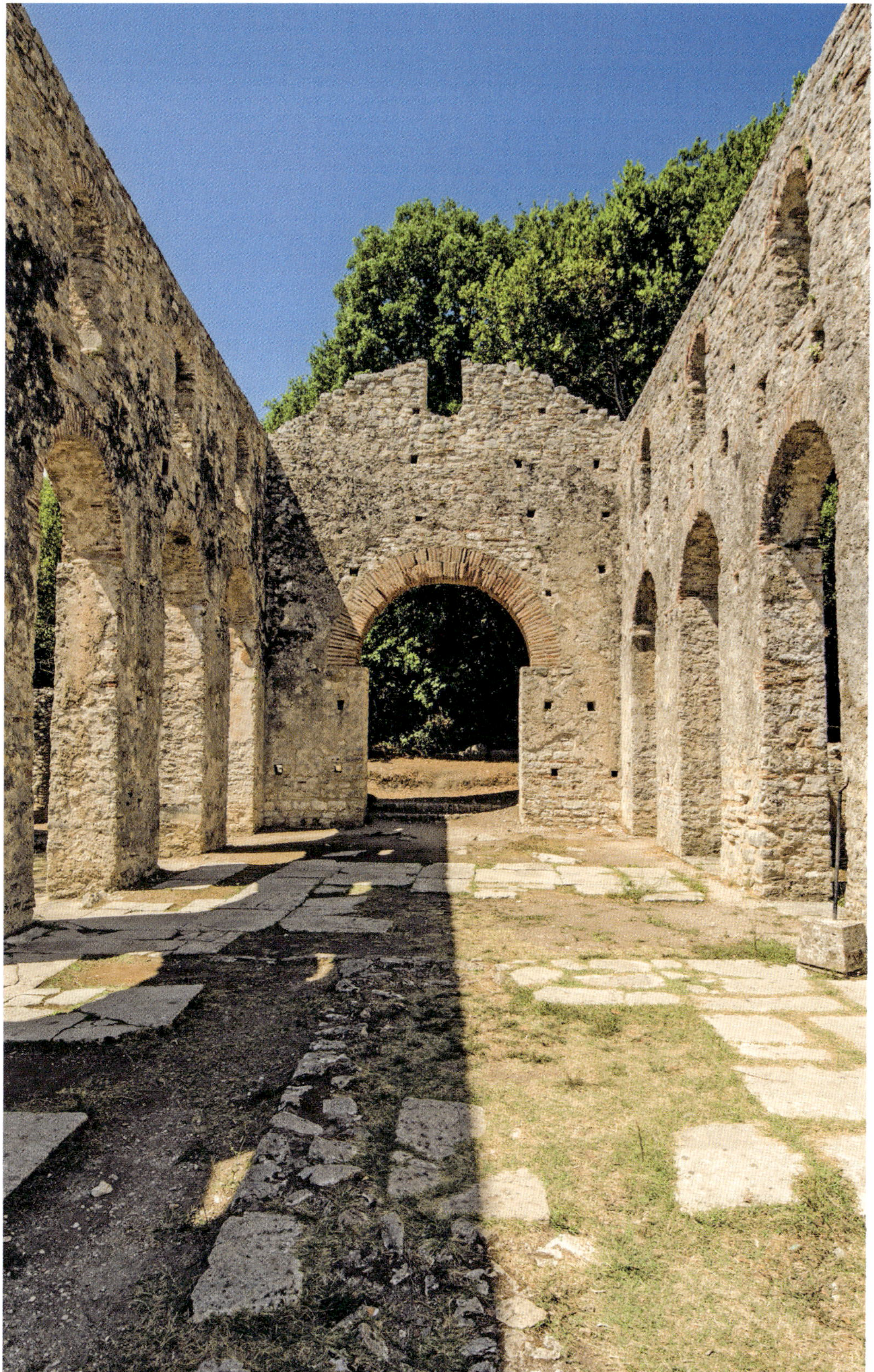

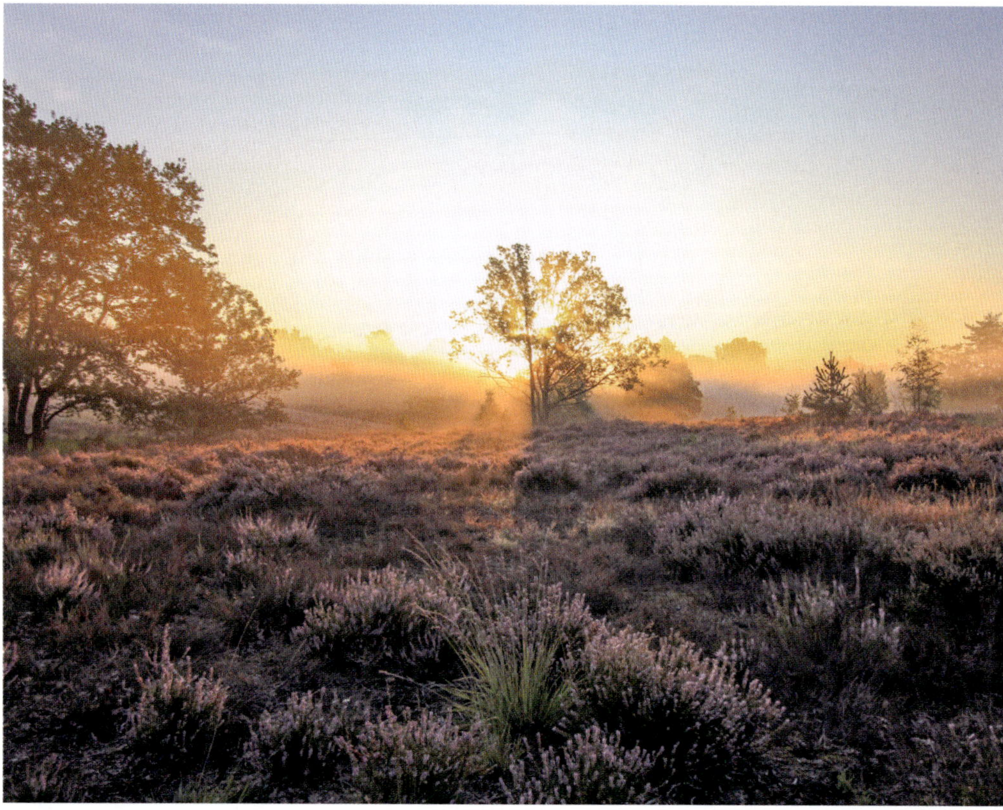

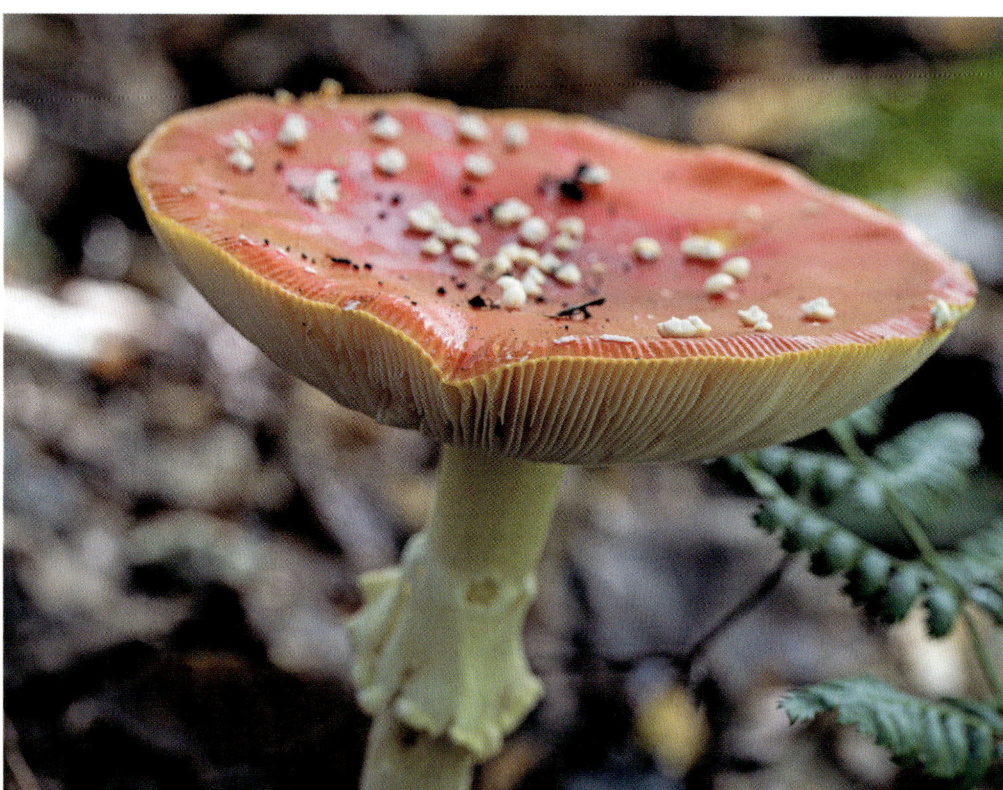

EUROPE   BELGIUM

109 HOGE KEMPEN NATIONAL PARK

TO VISIT BEFORE YOU DIE BECAUSE

An enchanting stretch of pine forests, dunes, and heathlands, Belgium's first and for a long time only national park is a great hiking and cycling spot.

Pint-sized Belgium has a handful of national parks, but this one, located on the eastern edge of the country, packs an outsized punch. The area stretches across 10 municipalities and encompasses pine forests, purple-flowered heathlands, dunes, marshes, and numerous ponds. Juniper, bell heather, and moss are plentiful, as are insects like dragonflies and silver studded blue butterflies. Wildlife includes moor frogs, natterjack toads, common lizards, red foxes, pine martens, and roe deer. More than 120 miles of trails weave through the protected land. And for those seeking a deeper understanding of the park's ecology and conservation efforts, the Hoge Kempen National Park Visitor Center provides a wealth of information through interactive exhibits, guided tours, and educational programs. Here, visitors can learn about the park's unique ecosystems, the threats they face, and the measures being taken to protect them for future generations.

www.nationaalparkhogekempen.be/en

EUROPE BULGARIA

110 RILA NATIONAL PARK

TO VISIT BEFORE YOU DIE BECAUSE

Some of Bulgaria's most beautiful mountain landscapes and ancient frescos are found in this large park in the southwest of the country.

As its name might suggest, this national park is dominated by the majestic Rila mountain range, which is not only the highest in Bulgaria but also home to the loftiest peak in the entire Balkan Peninsula – Musala Peak, standing at 9,598 feet above sea level. However, the park is also renowned for its glacial lakes, of which there are more than 120. The most famous among them are the mesmerizing Seven Rila Lakes, a cluster of seven stunningly beautiful, turquoise-colored lakes nestled amidst towering peaks and rocky cliffs. Also within the boundaries of the park lies the famous Rila Monastery, a Unesco World Heritage Site and one of Bulgaria's most revered cultural landmarks. Founded in the 10th century, the monastery still has vibrant frescoes and a rich collection of artifacts. For outdoor enthusiasts, Rila National Park offers many recreational opportunities year-round, from hiking and mountain biking along scenic trails in the summer to skiing in the winter months.

rilanationalpark.bg/en

EUROPE CROATIA

111 KRKA NATIONAL PARK

TO VISIT BEFORE YOU DIE BECAUSE

This Dalmatian gem boasts some of the most picturesque waterfalls in Europe, an island monastery and the ruins of medieval fortresses.

Krka National Park is a stunning natural oasis nestled in the heart of Croatia, renowned for its captivating waterfalls, limpid lakes, and lush greenery. Located along the Krka River in the Dalmatia region, this park's most prominent feature is its series of cascading waterfalls, the most famous being Skradinski Buk. Here, visitors can marvel at the sight of countless streams of water tumbling over limestone cliffs, cascading into a network of pools. Aside from its magnificent waterfalls, Krka National Park's other ecosystems include forests and lakes. Together, they support a variety of plant and animal species. There's a high number of birds of prey in particular, including the osprey, golden eagle, lanner falcon, peregrine falcon, and Eurasian eagle-owl, among others. For those interested in history and culture, the park is also home to several archaeological sites and cultural landmarks, including the medieval Visovac Monastery on Visovac Island. Founded in 1445, this sanctuary is found on a small islet in the middle of Visovac Lake, accessible only by boat.

www.npkrka.hr/en_US

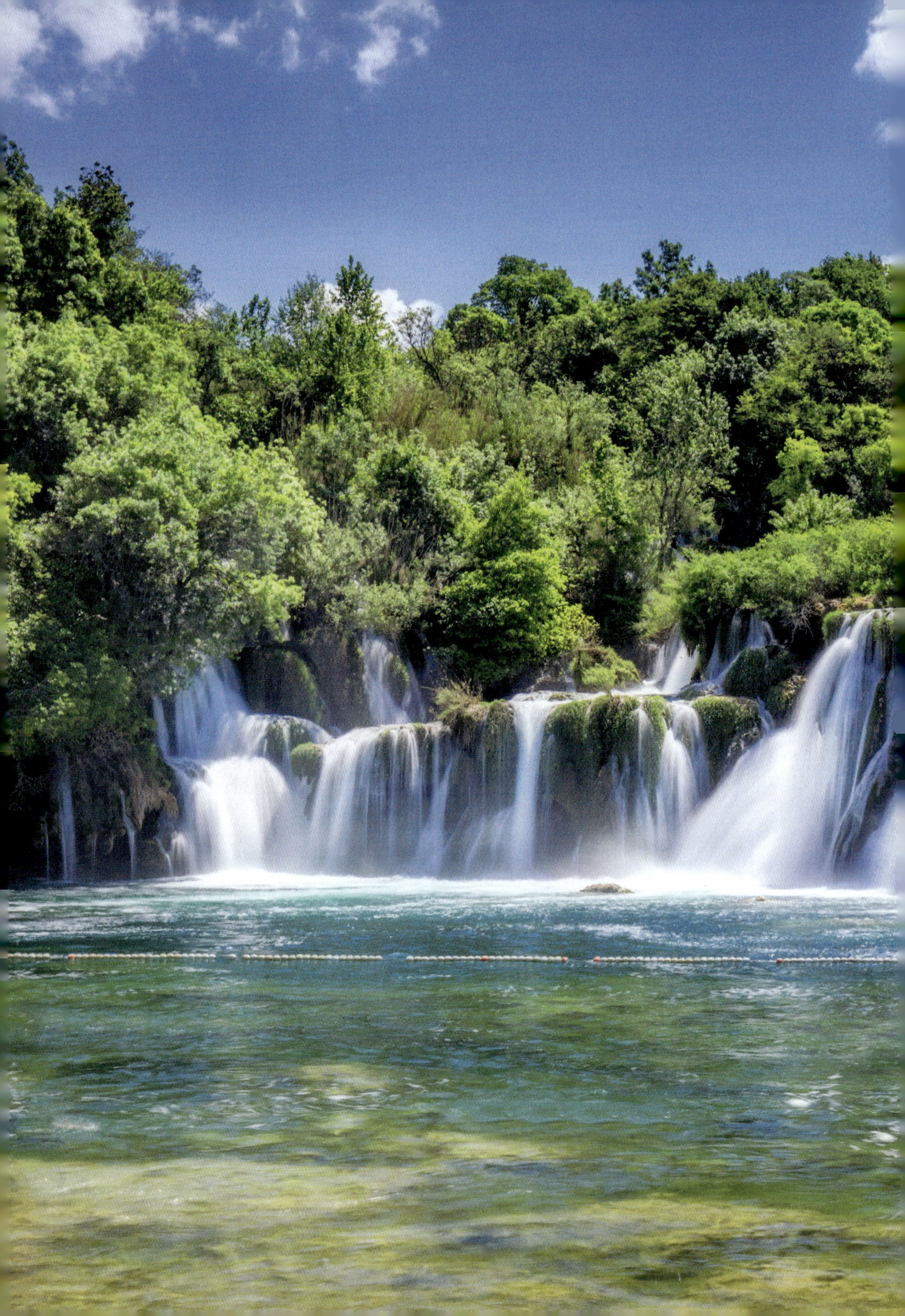

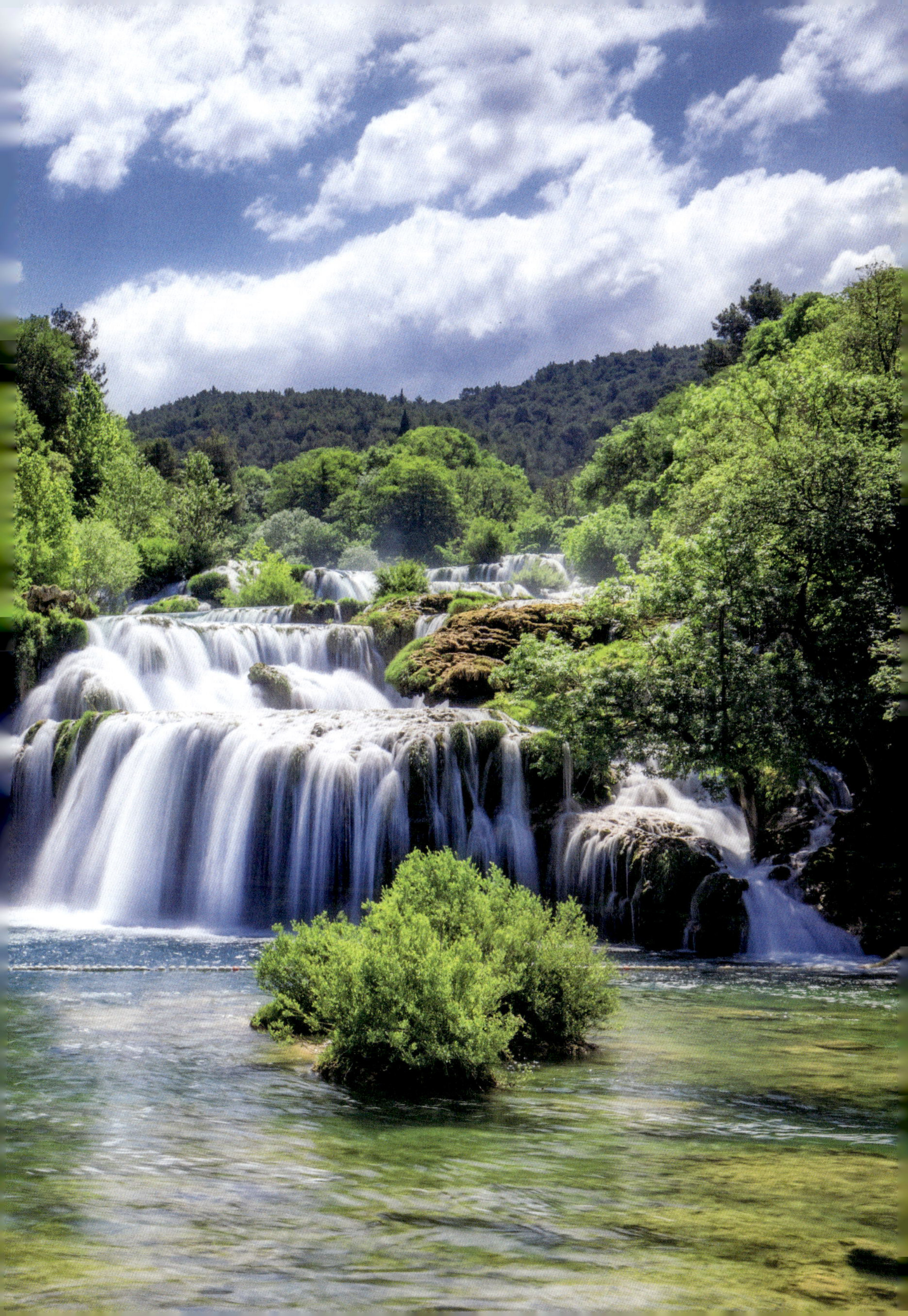

EUROPE   CROATIA

112 PLITVICE LAKES NATIONAL PARK

TO VISIT BEFORE YOU DIE BECAUSE

The largest national park in Croatia, it's known for its incredible, crystal-clear lakes and myriad waterfalls.

Covering roughly 115 square miles, Plitvice Lakes National Park is Croatia's largest reserve. It's known for its 16 brilliant cascading lakes, in vibrant shades of turquoise, azure, and emerald (though the colors frequently change depending on the minerals or organisms in the water and the angle of sunlight). These lakes, formed by the confluence of several small rivers and streams, are renowned for their remarkable transparency, allowing visitors to peer into their depths and observe the aquatic life below. The park's most iconic feature is undoubtedly its series of waterfalls ranging in size from gentle dribbles to thundering torrents plunging into the depths below. The most famous among them is Veliki Slap, or the Great Waterfall, which stands at an impressive height of more than 260 feet. Visitors to Plitvice Lakes National Park can explore these wonders via a network of well-maintained wooden boardwalks and hiking trails that wind through towering beech, fir, and spruce forests.

EUROPE FINLAND

113 HOSSA NATIONAL PARK

TO VISIT
BEFORE YOU DIE
BECAUSE

This park offers watersports, fishing, hiking trails, and ancient rock paintings in the remote, far northeastern region of Finland.

Hossa National Park is an unparalleled destination for outdoor adventures. Here, visitors can hike and cycle along 56 miles of trails through sprawling marshlands and taiga pine forests and go kayaking, canoeing, or fishing in any of the 130 crystal-clear lakes and ponds. Scuba diving at Öllön, a 131-foot-deep lake, is also possible. And when they tire, they can make a campfire to grill sausages in the dozen wilderness huts scattered throughout the protected land or pitch a tent at one of the designated campgrounds to spend the night. Also of interest are the 4,000-year-old rock paintings found in two canyons, Värikallio and Julma-Ölkky, which depict stories of shamanism (the latter canyon measures more than 2 miles long and 164 feet deep, making it the largest canyon in the country). Bears, wolverines, elk, wolves, and reindeer can be seen throughout the park. It's also beautiful in the winter, when activities include snowshoeing, cross-country skiing and ice-fishing.

www.nationalparks.fi/en-US/hossa

EUROPE • FINLAND

114 OULANKA NATIONAL PARK

TO VISIT BEFORE YOU DIE BECAUSE

No matter the season, there's something to do in this rugged and remote park, known for its river ecosystem and untouched boreal forest.

Located in Finnish Lapland, this park was named for the Oulanka River, a lifeline that meanders through the wilderness, carving its way through rocky gorges and lush valleys. With both tranquil stretches and challenging rapids, it's an exciting destination for kayakers, rafters, and anglers seeking adventure in Finland's vast wilderness. It's also a popular park for trekking – the famous 50-mile-long Karhunkierros route is located here. And in the winter, when the river freezes and the landscape is blanketed in snow, cross-country skiing, snowshoeing, and ice climbing are common activities, especially on the Rytisuo Snowshoeing Trail and the Oulanka Wilderness Trail. Wildlife populations thrive within the park, and visitors might be rewarded with sightings of brown bears, lynxes, and wolves on ground level or see golden eagles, ospreys, and capercaillies in the skies above. The local Sami people also still practice reindeer herding within the park, so chances are good you'll see them, too.

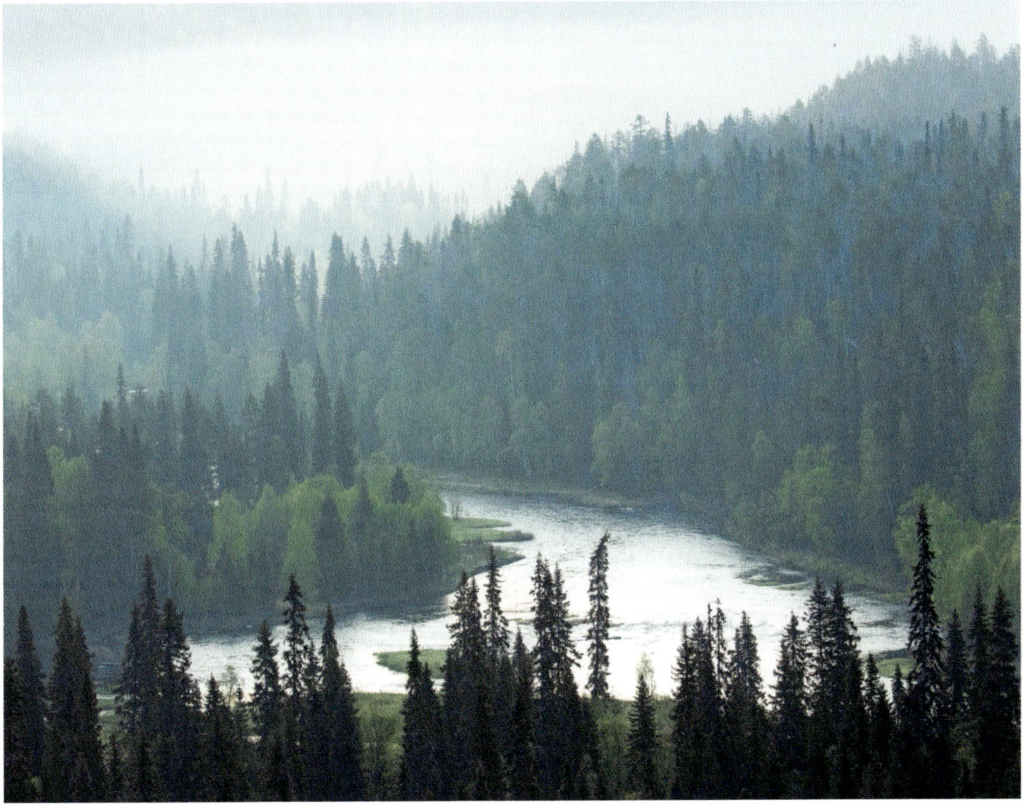

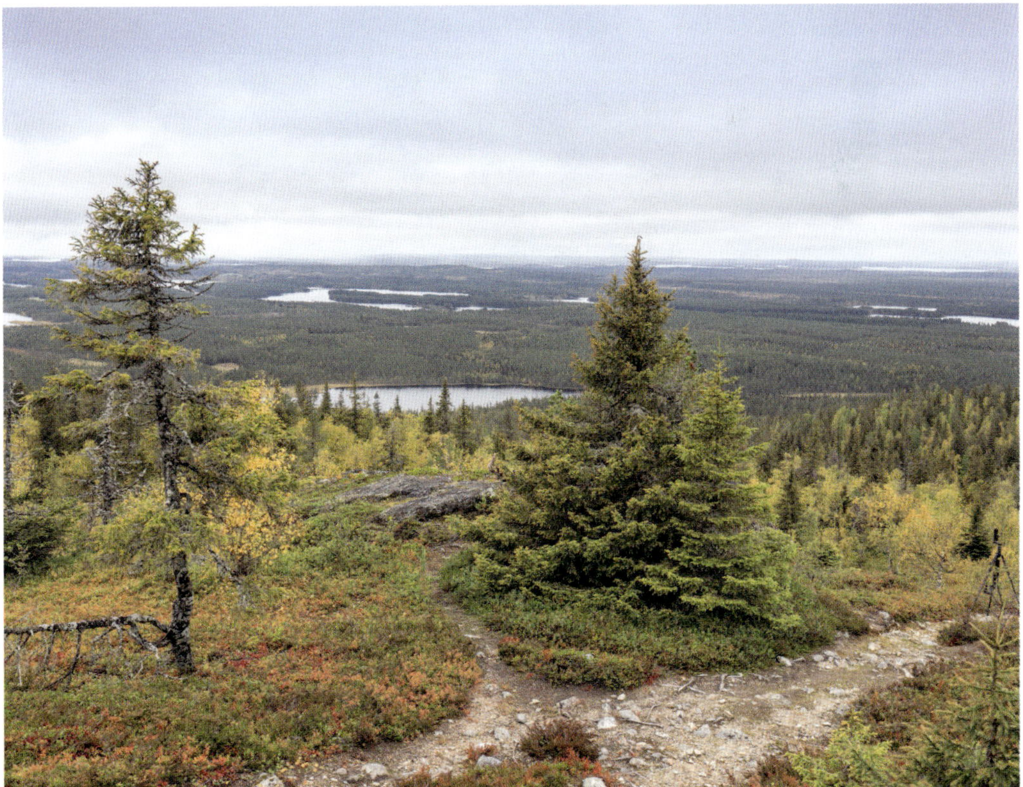

EUROPE GERMANY

115 SAXON SWITZERLAND NATIONAL PARK

TO VISIT BEFORE YOU DIE BECAUSE

With its bizarre rock formations and fascinating flora and fauna, this is one of the most enchanting hiking areas of Europe.

Despite its name, this national park is not located in Switzerland but rather in the eastern part of Germany, near the border with the Czech Republic. The park's defining feature is its striking sandstone formations – towering rock pillars, sheer cliffs, and deep ravines create a dramatic and awe-inspiring landscape. Myriad trails are found within Saxon Switzerland National Park, one of the most famous being the "Painters' Trail." This 72-mile route starts in the city of Pirna, follows the Elbe River east until Schmilka, and loops back to Pirna on the other side of the waterway. It offers incredible panoramic views of the Elbe River valley from the famed Bastei Bridge. One of the park's most iconic landmarks, the 76.5-meter stone bridge stretching between the towering rock formations, seems almost to blend in. The bridge itself is arguably one of the most photographed elements of the park.

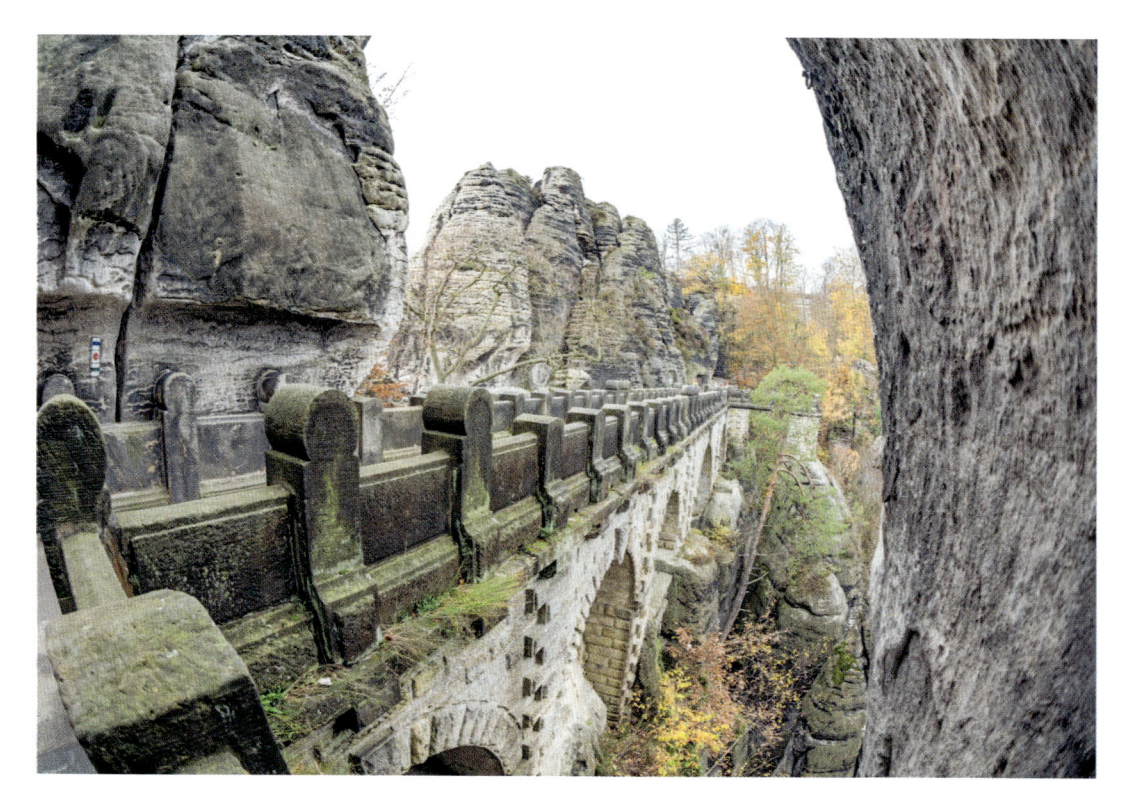

www.saechsische-schweiz.de/en

EUROPE GREECE

116 OLYMPUS NATIONAL PARK

TO VISIT BEFORE YOU DIE BECAUSE

Its namesake mountain inspired scores of myths and legends, while its gorges and forests are the setting of some incredible hikes.

In Greek mythology, Mount Olympus stands as the sacred domain of the ancient Greek gods. Its lofty peaks, wreathed in clouds and bathed in eternal sunlight, serve as the celestial court where the immortals convene to rule over the cosmos. In reality, Mount Olympus is the highest peak in Greece at 9,573 feet and part of a 92-square-mile national park of the same name. Given that its snow-capped summit often pierces the clouds, it's not hard to see why it was thought to be the mythical home of the gods. There are incredible hikes on and at the base of the fabled mountain. One popular option goes along Enipeas Gorge, crossing over wooden bridges and through dense forests, and passing the 15th-century Agios Dionysios Monastery before reaching a series of plunge pools and waterfalls. Canyoning along the Orlias Gorge and climbing on Xerolaki Ridge are other popular pursuits for adventurers.

olympusfd.gr/en

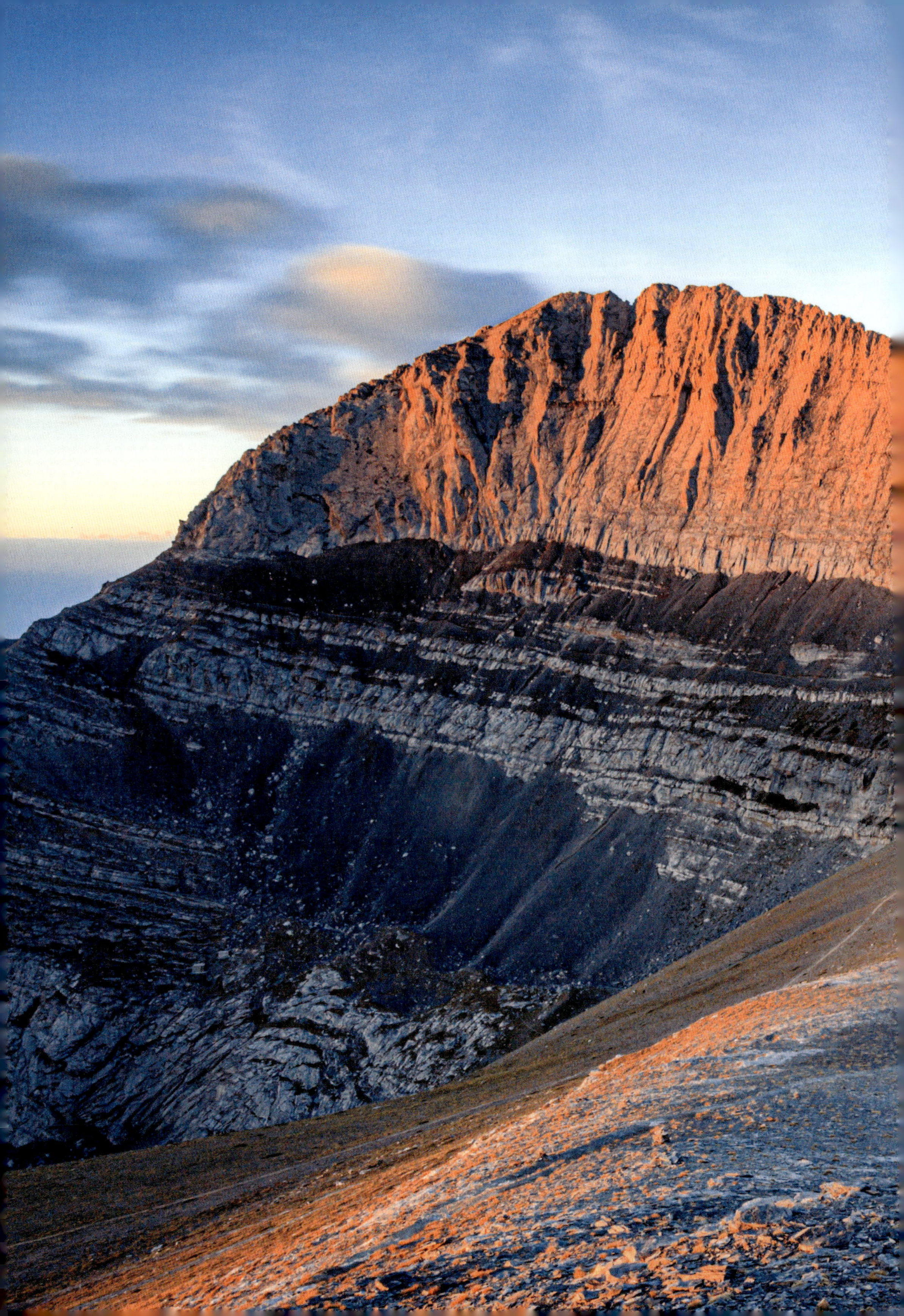

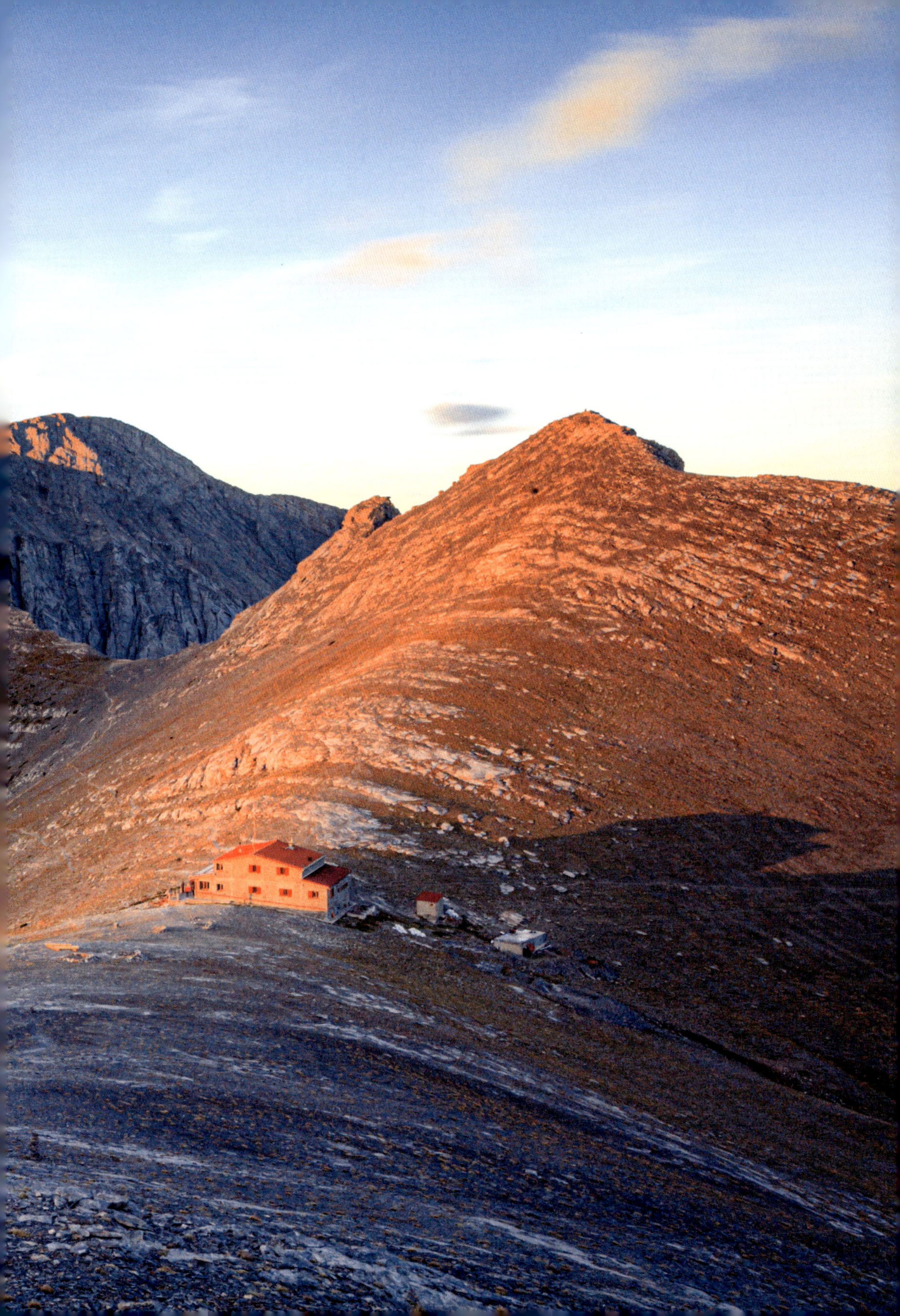

EUROPE GREENLAND (DENMARK)

117 NORTHEAST GREENLAND NATIONAL PARK

TO VISIT BEFORE YOU DIE BECAUSE

Head north to this immense protected area of the Greenland ice sheet for one-of-a-kind adventures in the Arctic.

Northeast Greenland National Park, established in 1974, stands as one of the world's largest protected areas, covering over 375,000 square miles of pristine Arctic wilderness (for context, that's an area larger than Tanzania, but smaller than Egypt). The centerpiece of the park is the Greenland ice sheet, the second-largest ice body in the world after Antarctica. The vast ice cap, which covers around 80 percent of Greenland's landmass, is responsible for sculpting the valleys and fjords of the island. It too is where the towering icebergs that drift along the coast come from. It's a land of extremes, where temperatures are below freezing much of the year and polar winds howl. Despite its harsh climate, the park teems with life, from polars bears and Arctic foxes to migratory birds that flock to its shores during the brief Arctic summer. For those willing to brave the elements, it's a park that offers a once-in-a-lifetime adventure.

EUROPE   ICELAND

118 VATNAJÖKULL NATIONAL PARK

TO VISIT BEFORE YOU DIE BECAUSE

It's one of just three national parks in Iceland and home to the largest non-Arctic glacier in Europe.

Centered around its namesake glacier, the spectacular Vatnajökull National Park is a wilderness area in southern Iceland. Spanning more than 5,000 square miles, the ice mass is generally about 1,300 to 2,000 feet thick (though it's estimated to be more than 3,000 feet in some areas) and blankets mountains, valleys, plateaus, and even active volcanoes. The park also includes Jökulsárlón, a milky glacial lagoon speckled with icebergs. Two famous waterfalls, Svartifoss and Dettifoss, are also within the boundaries of this protected area. For adventure-seekers there are myriad things to do within the national park, including solo or guided hikes (including some treks that are on the glacier itself), visiting the ice caves, going on a Zodiac boat tour, kayaking amongst the icebergs, horseback riding, participating in a snowboarding tour, rock climbing, camping, and more. There really is something for everyone.

www.vatnajokulsthjodgardur.is/en

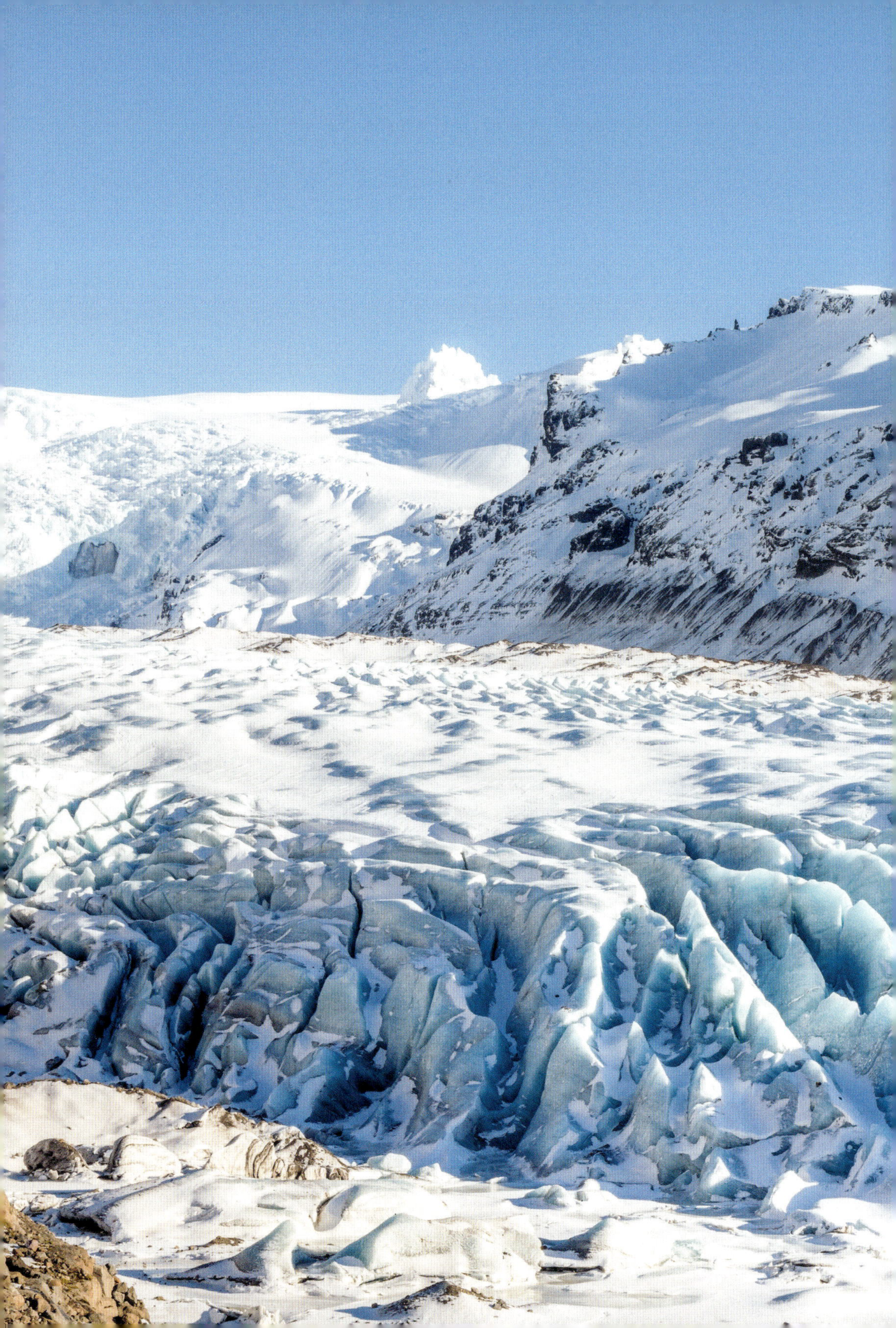

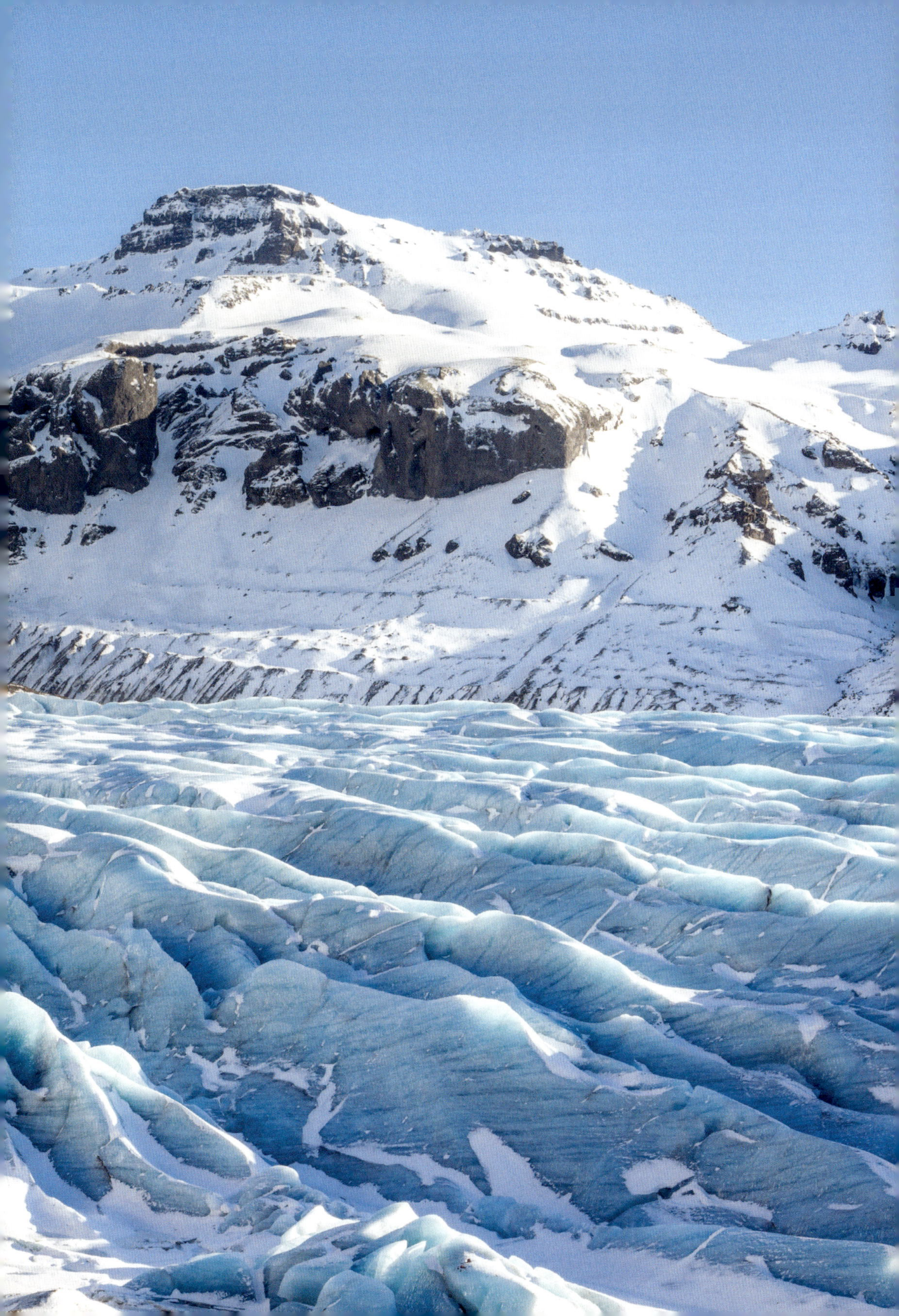

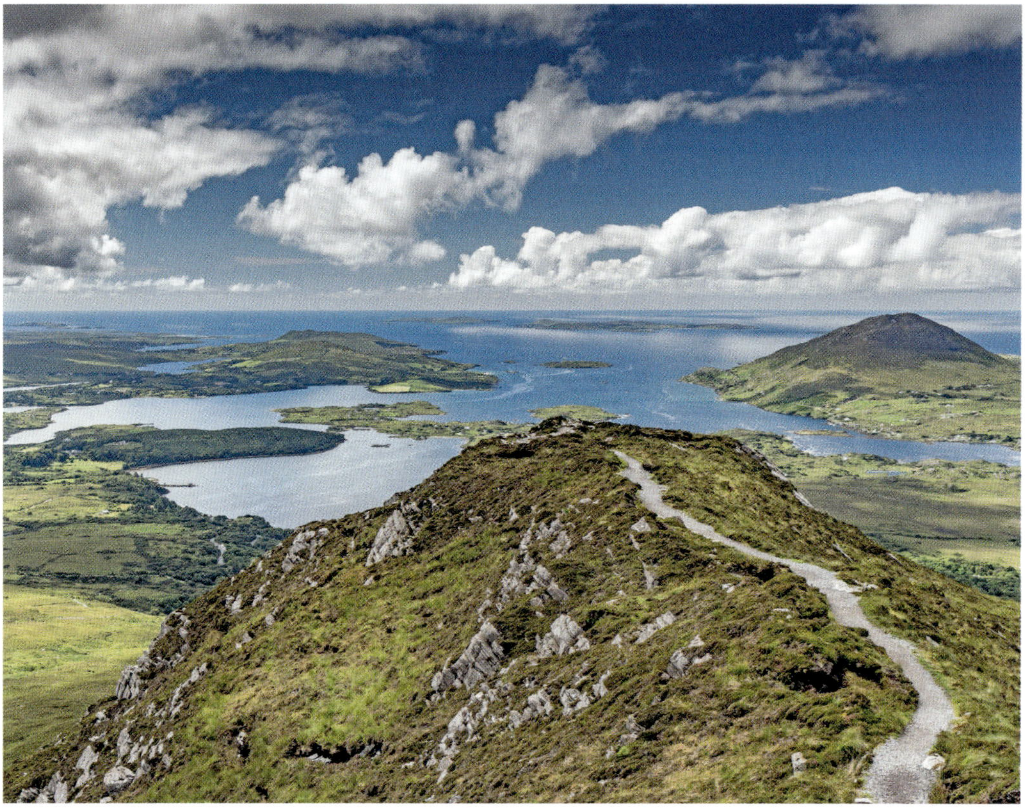

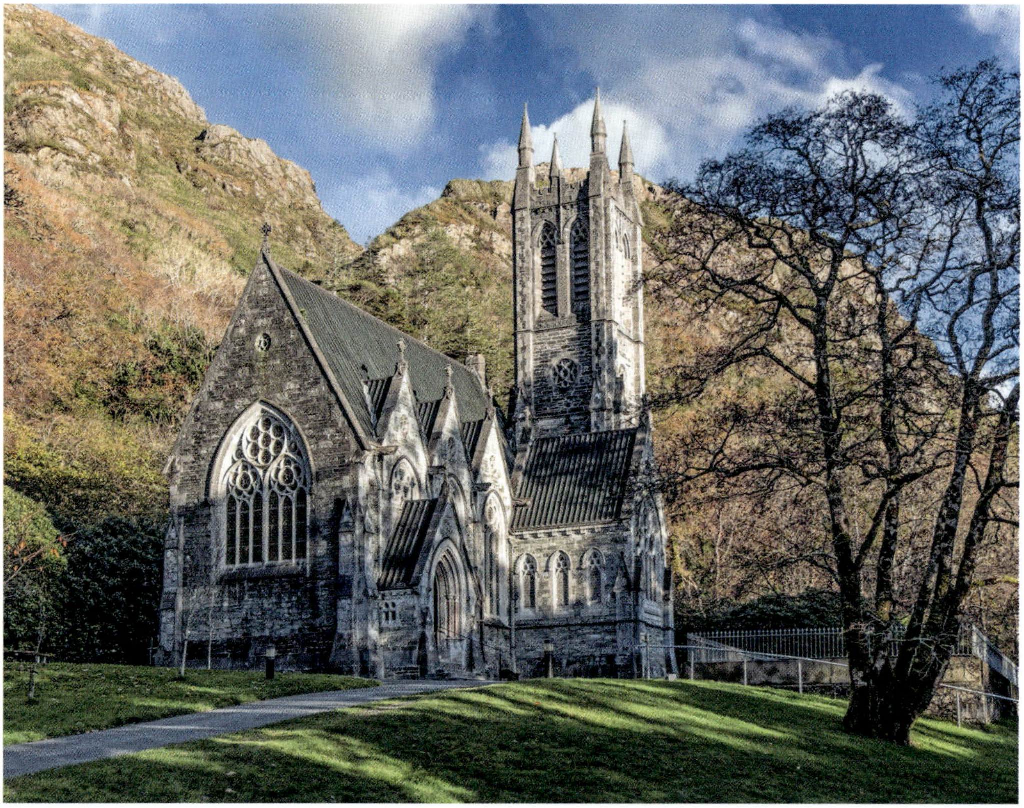

EUROPE   IRELAND

119 CONNEMARA NATIONAL PARK

TO VISIT BEFORE YOU DIE BECAUSE

Including parts of the Twelve Bens mountain range, this County Galway park is home to stunning hiking trails, Connemara ponies, and megalithic tombs.

Established in 1980, Connemara National Park is one of Ireland's six national parks and encompasses a diverse mix of habitats, including boglands, heathlands, mountains, and woodlands. Hiking is the primary activity in Connemara. There are just four established walking trails. One of the easiest is the one-mile Yellow Route, with its scenic views of Diamond Hill, Ballynakill Harbour, Barnaderg Bay, and Tully Mountain. For something more challenging, try the strenuous 2.3-mile Red Route, which takes trekkers to the 1,450-foot summit of Diamond Hill, where there are sweeping views of the Twelve Bens mountain range and Kylemore Abbey. Or, for those who have backcountry experience and gear, there are myriad opportunities for wild hiking. As you hike, keep an eye out for Connemara ponies, red deer, and peregrine falcons. To learn more about the area, pop into the park's visitor center in the old Letterfrack Industrial School, where informative exhibits detail the park's history, ecology, and conservation efforts.

EUROPE · ITALY

120 DOLOMITI BELLUNESI NATIONAL PARK

TO VISIT BEFORE YOU DIE BECAUSE

These famously beautiful mountains are ripe for adventure sports, including breathtaking via ferrata climbing paths for those not afraid of heights.

Home to a portion of the Dolomites, a Unesco World Heritage Site and the legendary mountain range in Europe's Eastern Alps, Dolomiti Bellunesi National Park exudes natural beauty. While not all of the Dolomites are here, some of the most dramatic limestone peaks are, including the mountains in the Alpi Feltrine, Monti del Sole, Pramper, Schiara, Spiz di Mezzodì, and Talvena ranges. Marmolada, the highest point in all of the Dolomites at just under 11,000 feet, is also within the park boundaries. Altogether, they provide a plethora of outdoor activities, including hiking, mountain biking, paragliding, rock climbing, via ferratas (Italian for "iron path"), and skiing for adventure-seekers. There are also emerald-colored lakes, hardwood and coniferous forests, and alpine meadows, which prove to be good habitats for animals like the golden eagle, the wolf, the wild cat, and the ibex. The park also offers botanical gardens, an environmental education center, and a nature museum for those looking for mellower pursuits.

www.dolomitipark.it

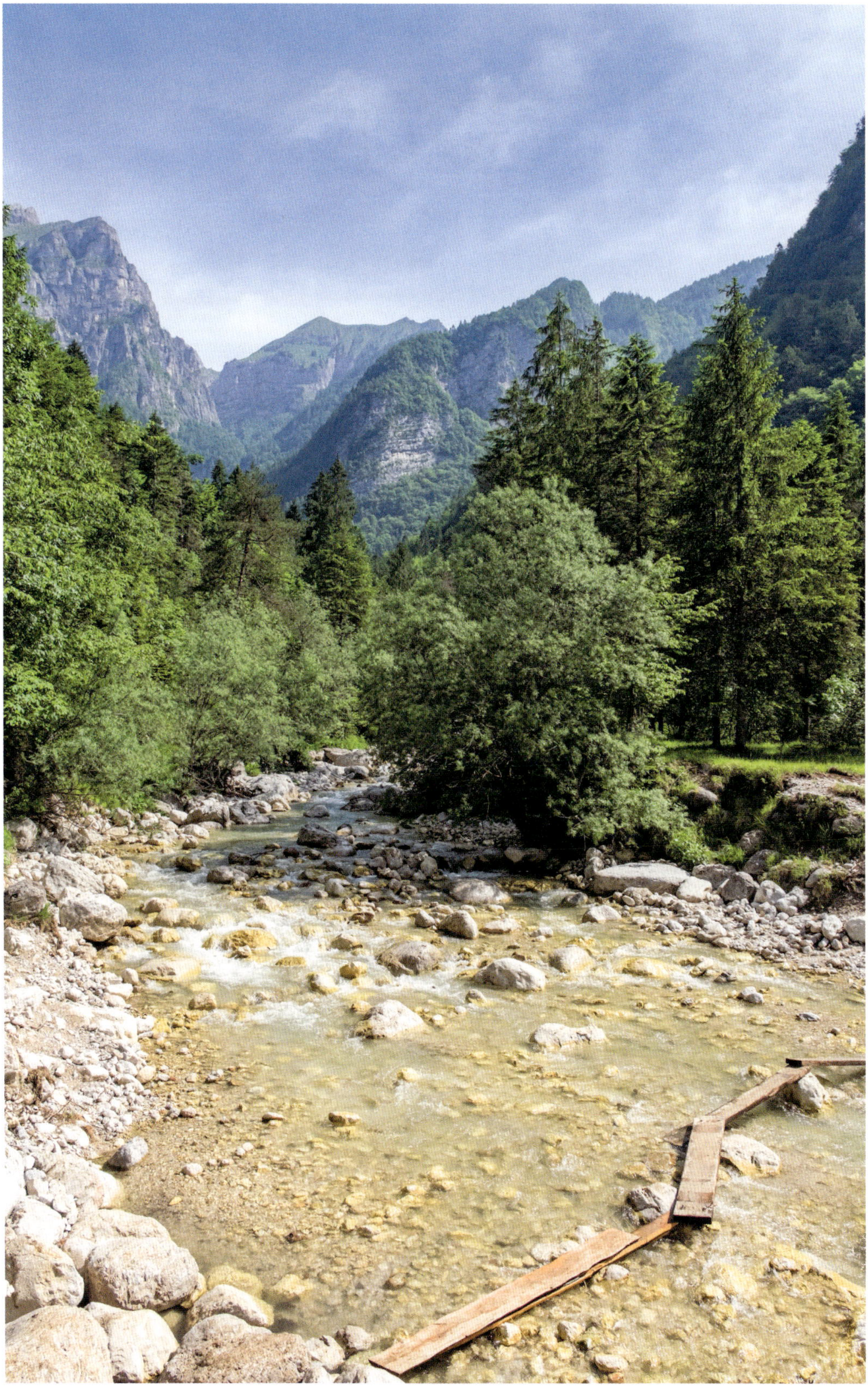

EUROPE　　　　　　　　　ITALY

121 GRAN PARADISO NATIONAL PARK

TO VISIT BEFORE YOU DIE BECAUSE

With herds of ibex clambering on its slopes, and pretty mountain villages, this Alpine park boasts great trekking and cross-country skiing.

Once the hunting grounds for the Savoys, the ruling house of Italy from 1861 to 1946, Gran Paradiso became the country's first national park in 1922 in an effort to protect the remaining ibex population, whose numbers had reached dangerously low levels (and have since rebounded). It is located between the Aosta Valley and Piedmont regions, and at its heart lies the towering Gran Paradiso massif, after which the park is named, reaching a staggering height of 13,323 feet above sea level. It, and other peaks such as Becca di Monciair and Ciarforon, offer technical challenges for those seeking adrenaline-fueled adventures. There are also more than 300 miles of trails at lower levels, including Giroparco Gran Paradiso and Alta Via Canavesana, which are more suitable to casual trekkers, bikers, or in the winter, cross-country skiers and snowshoers. The park also preserves a handful of villages with high mountain houses, Roman-era bridges, churches, and medieval castles, like Malgrà in Rivarolo Canavese.

www.pngp.it

EUROPE　　　　　　　　　ITALY

122 CINQUE TERRE NATIONAL PARK

TO VISIT BEFORE YOU DIE BECAUSE

Postcard-perfect villages, hikes through lemon groves, and swims in the sea – this crowd-pleasing coastal destination has it all.

Overlooking the azure waters of the Ligurian Sea, each of the five pastel-colored fishing villages of Cinque Terre (Corniglia, Manarola, Monterosso, Riomaggiore, and Vernazza) exudes its own distinct character and charm. It's possible to take the train between them, though there's also a network of scenic trails that connect them, meandering through terraced vineyards and fragrant citrus groves and offering breathtaking views of the coastline along the way. The famous Sentiero Azzurro, or Blue Trail, connects the five villages and provides an exhilarating trekking experience. For something more leisurely, visitors can also relax on sun-drenched beaches and take a dip in the crystal-clear waters or wander through the narrow cobblestone streets in search of artisan shops and trattorias. Cinque Terre is a paradise for food and wine enthusiasts, with its rich culinary tradition rooted in the freshest local ingredients. Sample the region's famous pesto, made from basil grown in the terraced gardens, savor freshly caught seafood, and indulge in the renowned Sciacchetrà, a sweet wine produced from grapes cultivated on the steep hillsides.

www.parconazionale5terre.it/Eindex.php

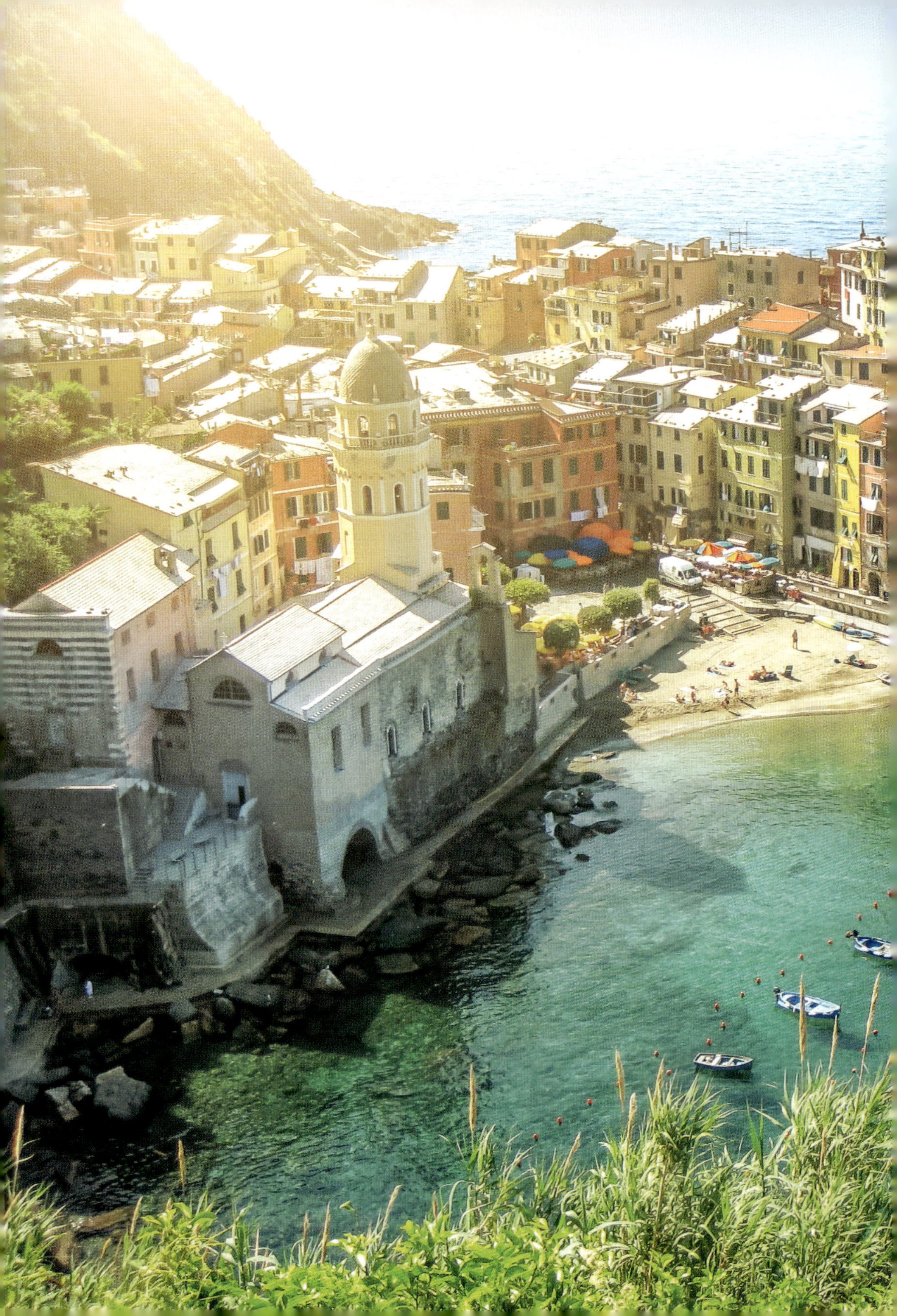

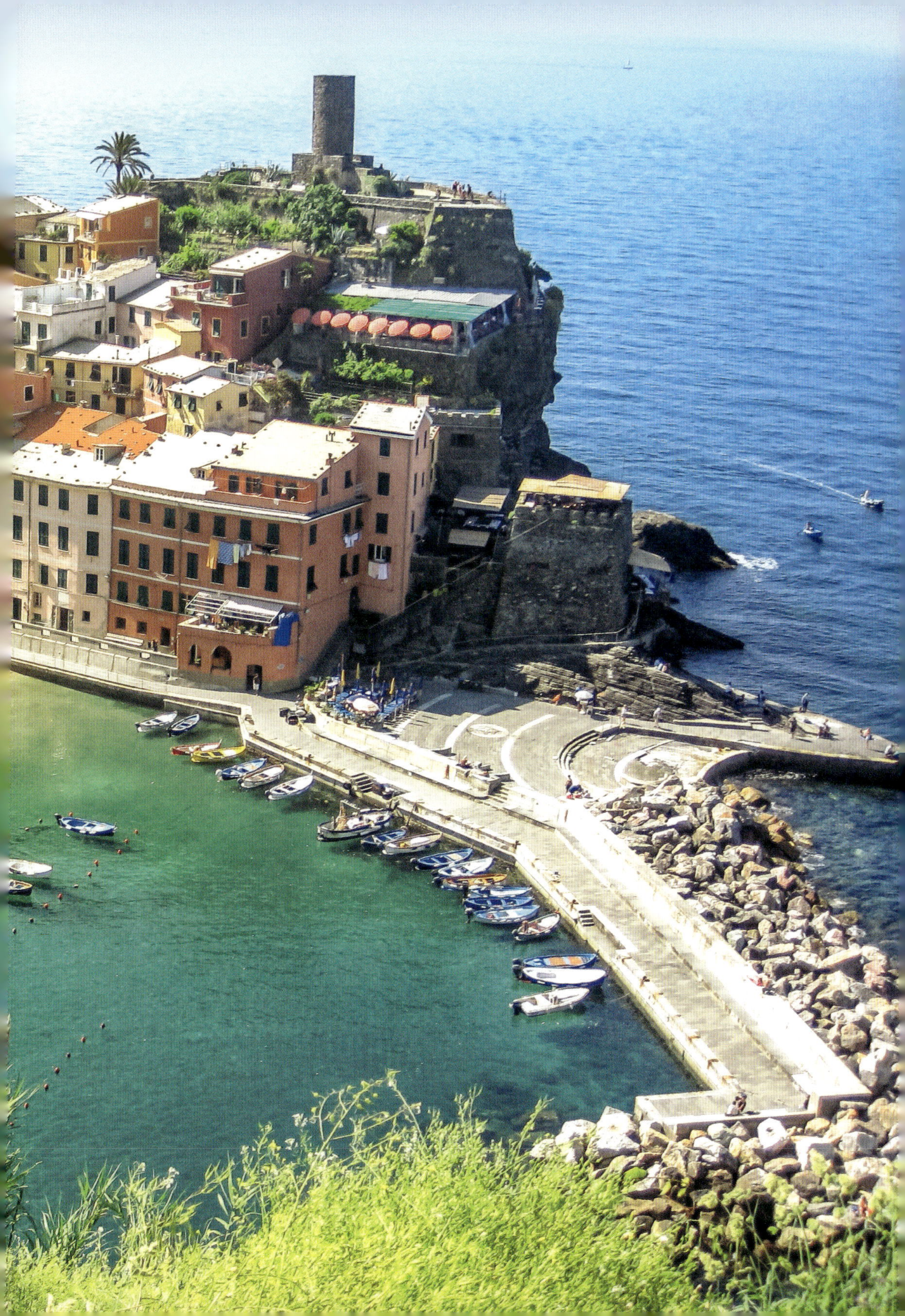

EUROPE LATVIA

123 GAUJA NATIONAL PARK

TO VISIT BEFORE YOU DIE BECAUSE

The largest and oldest national park in Latvia, it covers a large swath of the Gauja Valley and includes historic castles, towns, and museums.

Gauja National Park is a living museum of Latvia's rich heritage – there are more than 500 cultural and historical monuments within its territory. One of those sites is the Turaida Museum Reserve. Dating back to the 11th century, it's the most visited museum in Latvia and where visitors will find the imposing Turaida Castle, a manor house, and a church. Other castles include the Āraiši Lake Castle, a reconstruction of a 9th-century settlement on an island in Lake Āraiši, and the ruins of a Livonian Order castle in the medieval town of Sigulda. Cesis Old Town is another highlight – there, visitors can try their hands at handicrafts and games from the Middle Ages. There's also the Ungurmuiža Manor, the only wooden mansion from the Baroque era to survive to modern times in Latvia. Beyond the historical, Gauja has impressive biological diversity. One of the park's most striking features is the Gauja River Valley, carved out over millennia by the eponymous river. Here, visitors can marvel at towering sandstone cliffs, hidden caves, and tranquil stretches of water.

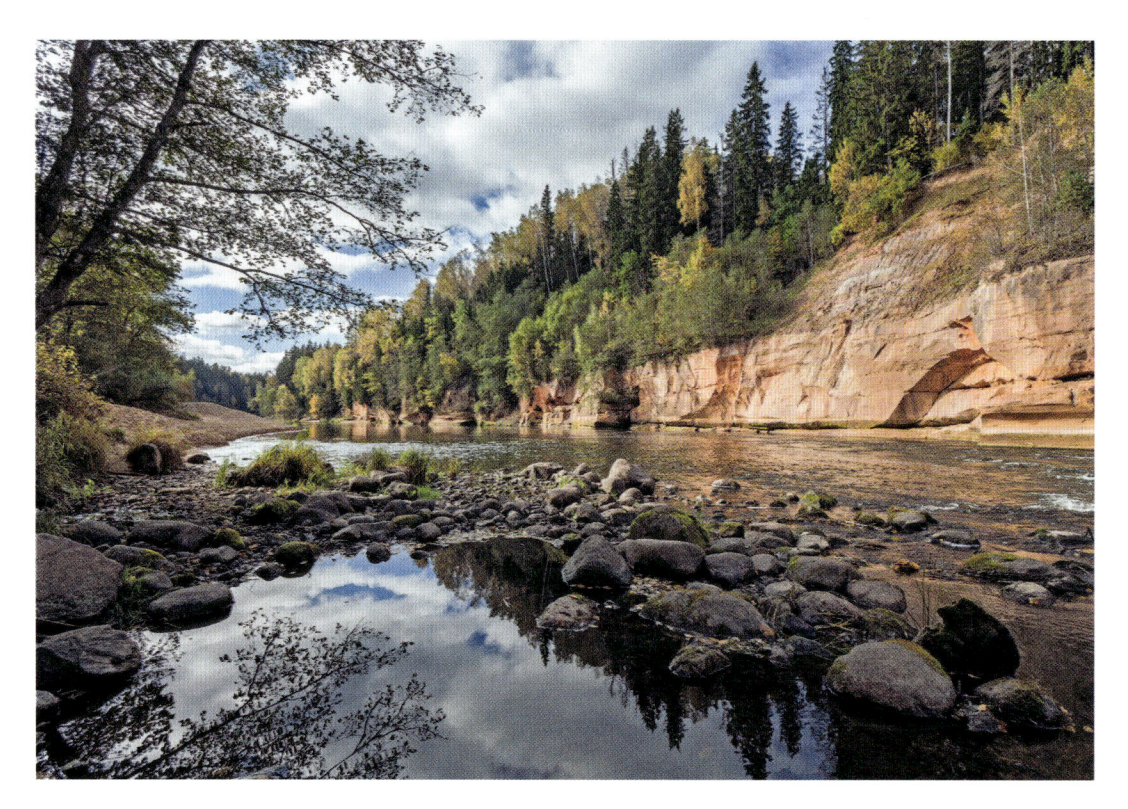

www.entergauja.com/lv

EUROPE　　　　　　　　　NORWAY

124 JOSTEDALSBREEN NATIONAL PARK

TO VISIT BEFORE YOU DIE BECAUSE

Glaciers make up nearly half of this national park – the other half is verdant valleys where you can picnic next to sparkling waterfalls.

If you've come to this Norwegian national park, it's likely to see its namesake glacier, the largest in mainland Europe. With a total glacial area of roughly 500 square miles, the otherworldly beautiful Jostedalsbreen glacier boasts an icy landscape adorned with crevasses, icefalls, and turquoise-blue meltwater lakes. It's that and other glaciers that have helped shape the surrounding landscape, giving it a terrain that ranges from deep valleys to towering peaks that pierce the sky. It's truly a testament to the dynamic forces of nature. However, the entire national park isn't just ice and rock – just hit the network of well-maintained hiking trails that wind through its lush valleys filled with wildflowers like purple saxifrage and mountain chickweed, and forested regions with elms and willow trees. Keep an eye out for deer, wild reindeer, lynx, wolverine, and the occasional bear (and birds like buzzards, white-backed woodpeckers, and golden eagles).

jostedalsbreen.no/en

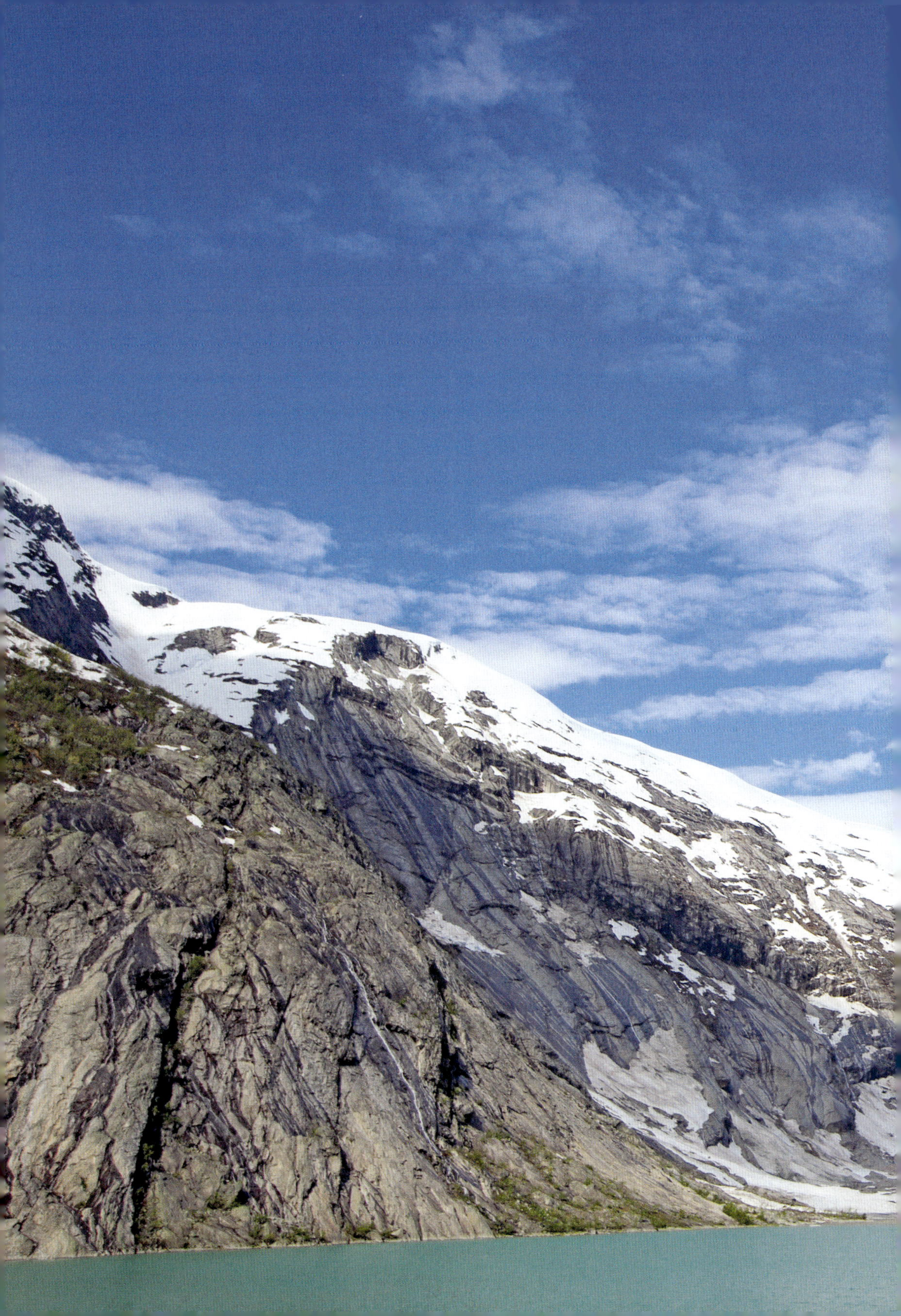

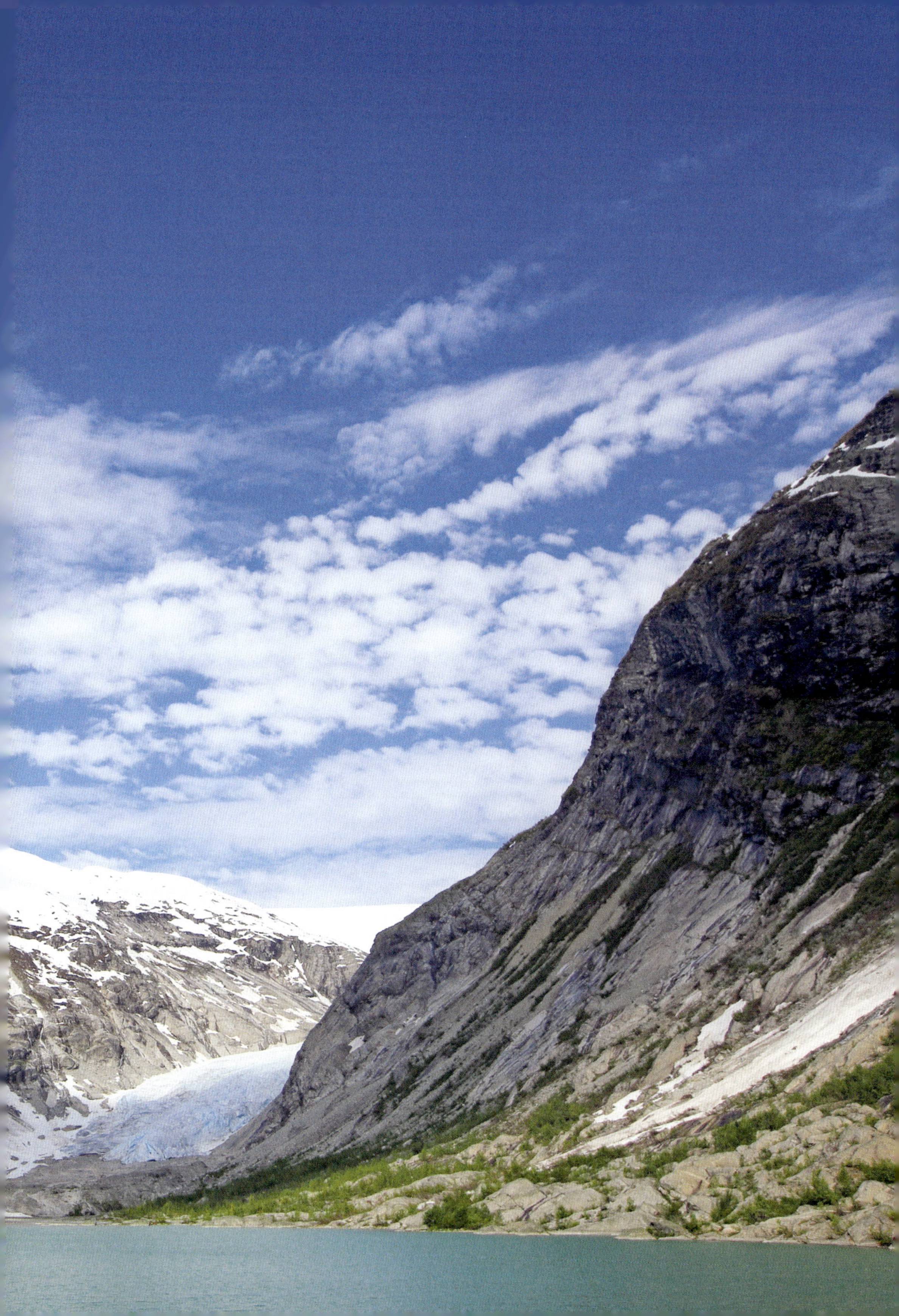

EUROPE　　　　　　　NORWAY

125 JOTUNHEIMEN NATIONAL PARK

TO VISIT BEFORE YOU DIE BECAUSE

With over 250 mountains, this scenic park north of Bergen offers outdoor adventures for all four seasons and all activity levels.

Translating to "Home of the Giants" in English (a moniker that was bestowed by Norwegian poet Aasmund Olavsson Vinje in 1862), Jotunheimen lives up to its name with towering peaks, deep valleys, and icy glaciers that dominate the landscape. Spanning over 1,000 square miles, Jotunheimen is Norway's largest national park and a haven for outdoor enthusiasts, particularly hikers, who can participate in everything from leisurely strolls to multi-day treks; and climbers, who are keen on tackling some of Norway's highest peaks, including Galdhøpiggen, the tallest mountain in the country. The park is also home to a via ferrata which allows novice climbers to ascend mountainous terrain with the help of metal rungs and footplates. In the winter, when deep snow blankets the area, ski touring is a popular local pursuit, as is off-piste snowboarding, snowshoeing, and going for dog sled rides.

jotunheimen.com

EUROPE   POLAND

126 TATRA NATIONAL PARK

TO VISIT BEFORE YOU DIE BECAUSE

Superlative waterfalls, caves, and peaks, plus rare animals such as golden eagles and lynxes, in the High Tatras of southern Poland.

Straddling the border of Poland and Slovakia, this pristine wilderness sanctuary beckons nature lovers and adventure-seekers alike. Established in 1954 to safeguard the jagged peaks, spruce forests, deep glacial valleys peppered with wildflowers, and crystal-clear alpine lakes of the region, it is one of the oldest national parks in Central Europe. Rare and endangered animals find refuge within the borders of this national park, including the Tatra chamois, gray wolf, European otter, Eurasian lynx, brown bear, and golden eagle. For hikers and mountaineers, the biggest draw is Rysy, the tallest peak in the park at 8,199 feet. Visitors can also visit Jaskinia Wielka Śnieżna, the longest of the 600 caves found here. It stretches more than 11 miles in length and reaches a maximum depth of 2,671 feet. Also compelling is Wielka Siklawa, the tallest waterfall in the area, which cascades 230 feet.

zpppn.pl/tatra-national-park-en/park

EUROPE PORTUGAL

127 PENEDA-GERÊS NATIONAL PARK

TO VISIT BEFORE YOU DIE BECAUSE

The first and only national park in Portugal, this beautiful mountainous area is home to Garrano horses, remote hamlets, and historic sites.

Located in the northwestern corner of Portugal, Peneda-Gerês National Park (or simply Gerês) covers more than 270 square miles and protects the area's ancient history. Here, visitors will find megalithic stone tombs from the 3rd century, Celtic fortifications, Roman roads and milestone markers, a monastery from the 9th century (Santa Maria dos Pitões), and two castles that are roughly 1,000 years old (Castro Laboreiro and Castelo de Lindoso). There are still some remote hamlets, with their stone houses and narrow cobblestone streets, where shepherds tend to their livestock. Perhaps the park's most famous inhabitants today are the wild Garrano horses, who have been native to the region since the first millennium. And some of Europe's few wolves still roam within the park's oak forests, peat bogs, and bushlands. There are also foxes, ibexes, wild boar, deer, otters, and more than a dozen bat species (many of which are endangered).

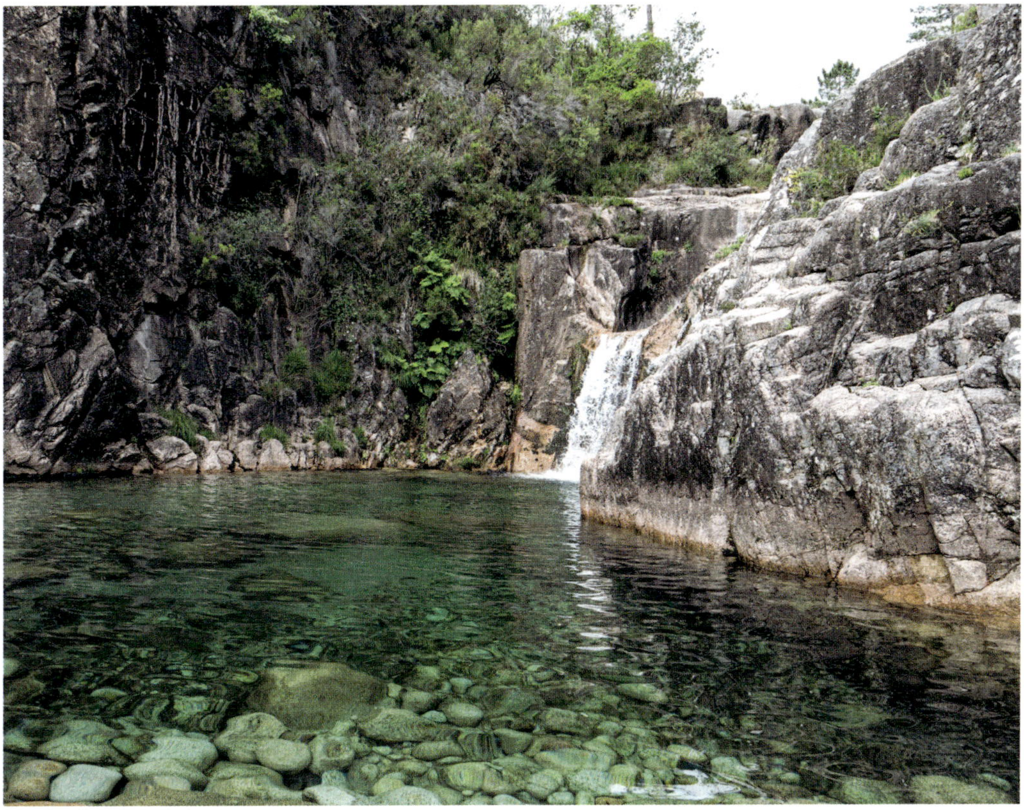

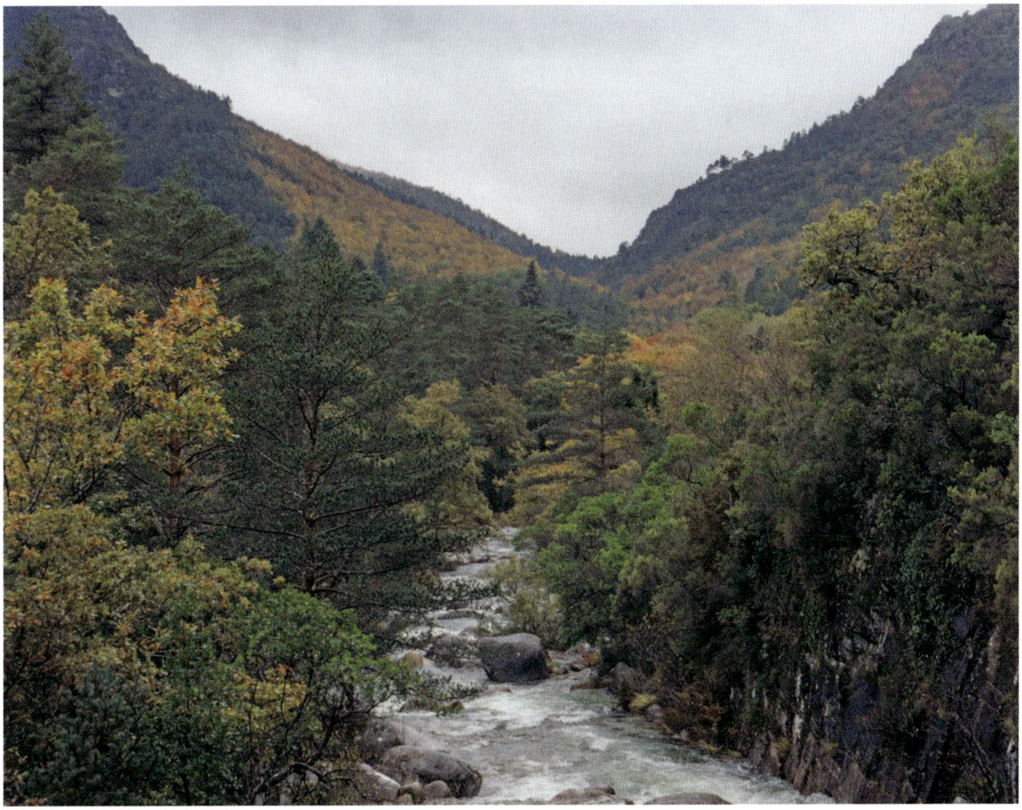

EUROPE   ROMANIA

128 CHEILE NEREI-BEUSNITA NATIONAL PARK

TO VISIT BEFORE YOU DIE BECAUSE

Head to the Carpathian Mountains to see the longest gorge in Romania, a unique mossy waterfall, and a pair of intriguing lakes, source of many legends.

There are 14 national parks in Romania, though arguably none are more beautiful than Cheile Nerei-Beusnita National Park. One of the most iconic features of this park, located in the Carpathian Mountains, is the Beușnița Waterfall, a splendid cascade that plunges over 65 feet into a turquoise pool below. Surrounded by lush vegetation and rugged cliffs, this natural marvel is a photographer's dream and a popular destination for nature lovers seeking serenity. Other waterfalls not to be missed are Văioaga and Bigăr. Equally mesmerizing is the Nera Gorge, a dramatic limestone canyon carved into existence by the meandering Nera River over the course of millions of years. On either end of the gorge visitors will find two karstic lakes, Ochiul Beiului (translated as the Bey's Eye) and Lacul Dracului (meaning Devil's Lake), both of which are featured prominently in local folklore.

www.romania.travel/en

EUROPE SCOTLAND (UK)

129 CAIRNGORMS NATIONAL PARK

TO VISIT BEFORE YOU DIE BECAUSE

This Highland park features rugged mountains, ancient Caledonian pine forests, sparkling lochs, and expansive moorlands.

Located in the heart of the Scottish Highlands, Cairngorms National Park is the largest national park in the United Kingdom. The protected land was named after the nearby Cairngorm Mountains, which dominate the skyline with their imposing peaks – five of Scotland's six highest mountains lie within the park. The landscape is dotted with numerous lochs and rivers, providing ample opportunities for fishing, kayaking, and wildlife spotting. The River Spey, famous for its salmon fishing, flows through the park, while Loch Morlich and Loch an Eilein offer serene settings for picnics and leisurely walks. Though the park contains a quarter of the United Kingdom's threatened species, it's good for wildlife viewing. On land, you may find pine martens, Scottish wildcats, red squirrels, and deer. Or if you turn your eyes to the skies, it is possible to see some of the golden eagles, ospreys, capercaillie, ptarmigans, and Scottish crossbills that call the Cairngorms home.

cairngorms.co.uk

EUROPE · SLOVAKIA

130 SLOVAK KARST NATIONAL PARK

TO VISIT
BEFORE YOU DIE
BECAUSE

A vast network of caves makes this park in southeast Slovakia a subterranean wonderland.

It should come as no surprise that this national park holds one of the largest karstic areas in Europe. These unique limestone formations, sculpted over millions of years by the erosive forces of water, have created a mesmerizing landscape of sinkholes, gorges, canyons, and limestone pavements. It is estimated that there are more than 1,100 underground chambers across the park and its twin park on the other side of the Hungarian border, Aggtelek Karst. Among the most famous formations at Slovak Karst National Park are the Domica Cave and Jasovská Cave, both of which offer visitors the chance to explore their subterranean wonders. Guided cave tours are a popular way for visitors to experience the park. Above ground, there are more than 300 miles of cycling trails, some of which lead to other must-see viewpoints, including Zádiel Gorge and the waterfalls of Háj.

slovakia.travel/en/national-park-of-slovensky-kras

EUROPE · SLOVENIA

131 TRIGLAV NATIONAL PARK

TO VISIT BEFORE YOU DIE BECAUSE

Climb Slovenia's iconic Mount Triglav and the surrounding high mountain pastures, forests, and valleys home to bears and lynxes.

There's a saying that claims you are not a true Slovenian until you climb to the summit of Mount Triglav, the peak after which this park is named. There are few things that elicit the same level of national pride for Slovenians than this towering massif that is often called the symbol of the country. However, you don't need to be Slovenian to climb it – or even a hardcore climber, for that matter. There is a slew of routes with varying degrees of difficulty to the top and even inexperienced trekkers can get atop it with the help of a guide. Or if that's not your jam, there are oodles of hikes at lower elevations, many of which cut through alpine meadows bursting with delicate wildflowers or follow crystal-clear streams that are fed by glacial runoff. Afterward there are a handful of excellent lodges to have a post-hike beer and apple pie at. There are approximately 7,000 different animal species living here, including ibexes, marmots, pygmy owls, salamanders, bears and lynxes.

EUROPE SPAIN

132 ORDESA AND MONTE PERDIDO NATIONAL PARK

TO VISIT BEFORE YOU DIE BECAUSE

Staggeringly beautiful valleys and peaks formed over thousands of years form a dramatic panorama at this park in the Aragonese Pyrenees.

Meaning "Lost Mountain," Monte Perdido stands tall at 11,000 feet above sea level, earning its name because the summit often seems to vanish into the clouds. The third tallest mountain in the Pyrenees, it's the centerpiece of this Spanish national park. The Ordesa Valley, to which the other part of the park's name is attributed, was carved by glaciers over millennia, and now the Arazas River meanders through its verdant expanse. The most impressive views of the mountain are seen from this valley. Wildlife of Ordesa and Monte Perdido National Park includes brown bear, boar, red squirrel, marmot, red fox, and badger, as well as various raptor species, like the Egyptian vulture, the bearded vulture, and the golden eagle. With an array of hiking trails within the park, from leisurely strolls along the valley floor to challenging ascents to high-altitude viewpoints, there's a route for every level of experience.

www.spain.info/en/nature/ordesa-monte-perdido-national-park

EUROPE SWEDEN

133 ABISKO NATIONAL PARK

TO VISIT
BEFORE YOU DIE
BECAUSE

Located in the heart of Swedish Lapland, this 7,700-hectare park offers the ideal conditions for seeing the northern lights.

Located about 150 miles north of the Arctic Circle, Abisko National Park is perfectly situated under the auroral oval, a band hugging the northernmost climes, where the northern lights are largely concentrated. From October until the end of March, as long as there isn't cloud cover, the chances are very good you'll see kaleidoscopic ribbons of light dance across the sky. Luckily, Abisko is home to what scientists call a "blue hole" – a weather pattern that consistently gives the area clear nights. That's not to say that all the excitement is after dark (or even limited to winter), though. Visitors can hike, snowshoe, or cross-country ski through Arctic birch forests, go dog sledding over frozen lakes, test their strength ice climbing, and learn about the Indigenous Sami people and their herds of reindeer.

EUROPE · THE NETHERLANDS

134 NATIONAAL PARK DE BIESBOSCH

TO VISIT BEFORE YOU DIE BECAUSE

This tranquil park near Dordrecht has one of the few freshwater tidal areas in the world, as well as willow forests and wet grasslands.

An intricate network of rivers, creeks, and marshes, this stunning landscape was created when 300 square kilometers of polder lands were submerged in the St. Elizabeth floods in 1421. The park is home to a rich variety of bird species, especially waterfowl that come here to forage for food, including herons, kingfishers, and spoonbills, making it a paradise for birdwatchers and nature enthusiasts. In addition to its avian inhabitants, De Biesbosch's marshy terrain is also inhabited by beavers, deer, and numerous species of fish and amphibians. Visitors to the park can explore its meandering waterways by boat, kayak, or canoe. Dozens of well-kept bicycle and hiking paths also weave throughout the park. There's also an excellent museum, the Biesbosch MuseumEiland, which offers displays on wildlife and the area's boating history. It's also unique in that it looks like it's built into a hill when in reality, its slated rooftop is covered in grass. It is possible to climb to the top of the roof for 360-degree views.

EUROPE　　　　　　　　THE NETHERLANDS

135 HET NATIONALE PARK DE HOGE VELUWE

TO VISIT BEFORE YOU DIE BECAUSE

A perfect blend of art and nature, this privately owned national park is home to a star-studded museum and stunning landscapes.

In the heart of the Netherlands, this natural preserve is characterized by its vast heathlands, dense oak and beech forests, and expansive sand dunes. It was the second national park in the Netherlands. Among its most well-known features is the Kröller-Müller Museum, which is surrounded by an outdoor sculpture garden and houses an impressive collection of art inside, including works by Vincent van Gogh, Pablo Picasso, Odilon Redon, Georges Seurat, Auguste Rodin, and Piet Mondrian. A second museum, Museonder, focusing more on the geology and biology of the area, is located in the park's visitor center. De Hoge Veluwe offers an extensive network of hiking and cycling trails (and a bicycle-sharing system that makes bikes free for visitors to use), allowing guests to explore its varied landscapes at their own pace. There's also a small wildlife population consisting of red deer, wild boar, foxes, badgers, European pine martens, and a pack of Eurasian wolves.

www.hogeveluwe.nl/en

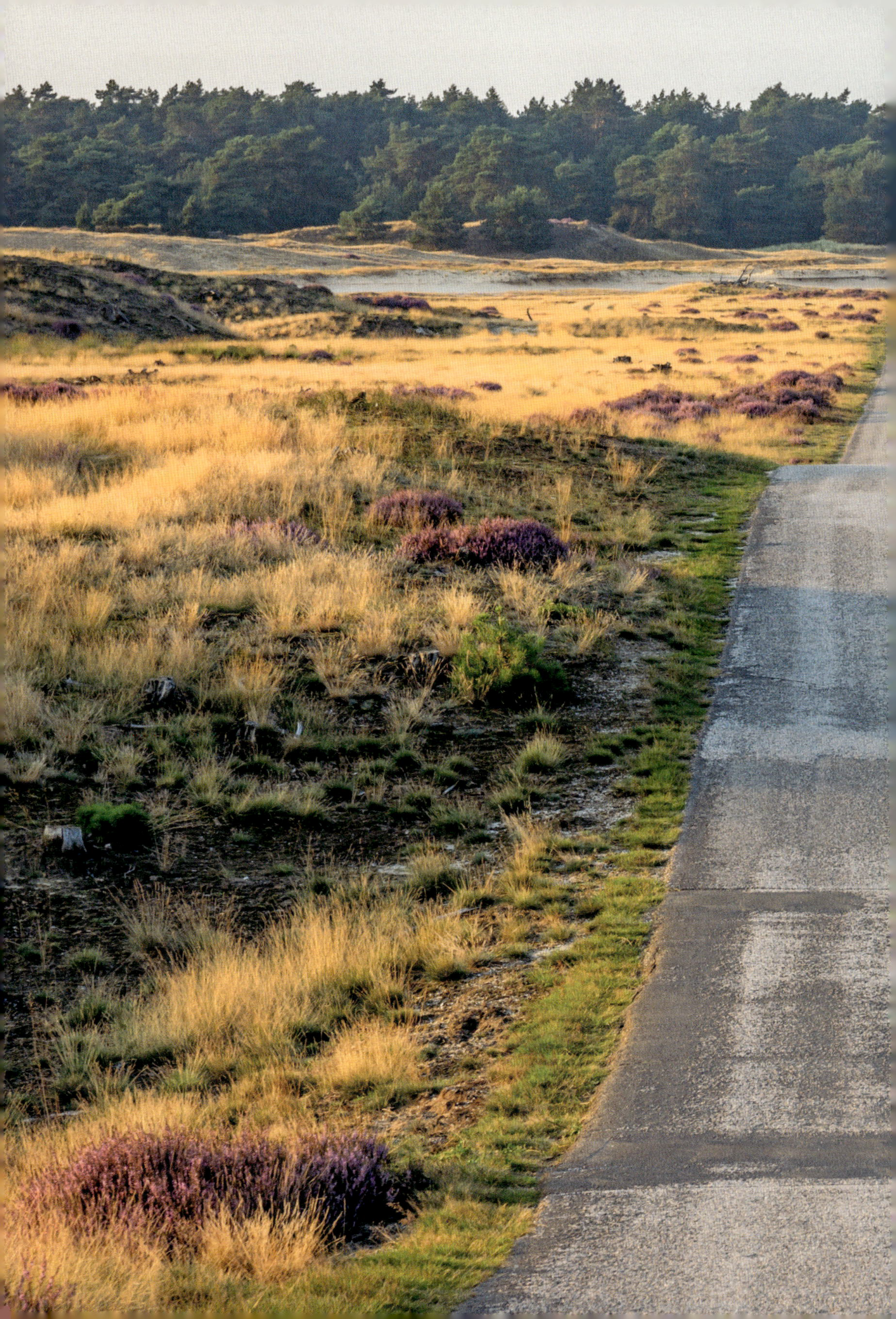

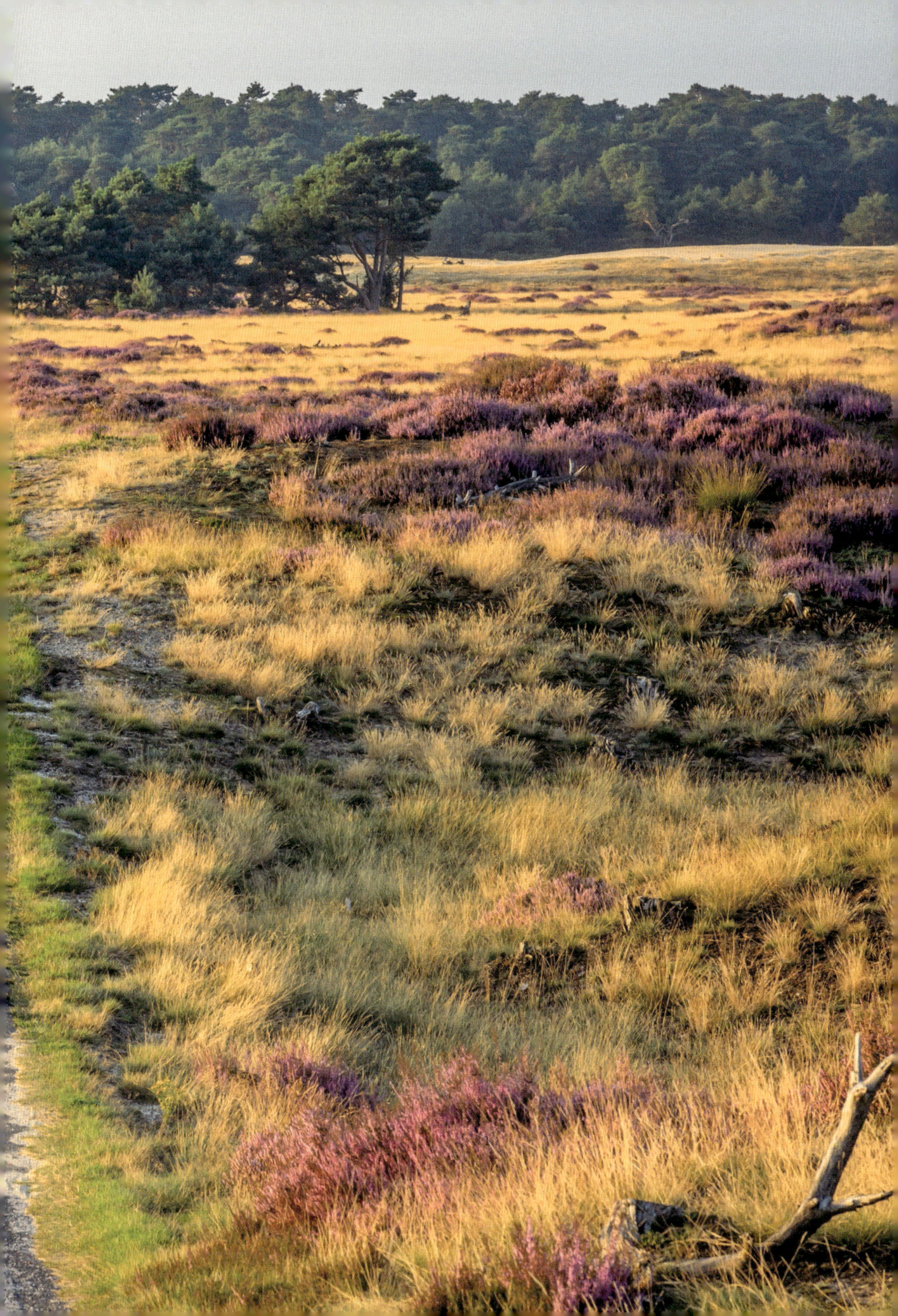

EUROPE THE NETHERLANDS

136 DE MEINWEG NATIONAL PARK

TO VISIT BEFORE YOU DIE BECAUSE

This beautiful terraced landscape spread on three different plateaus is a highlight of the low countries.

De Meinweg National Park, nestled in the southern province of Limburg in the Netherlands, is a picturesque park with a landscape characterized by terraces, stream valleys, rolling hills, dense woodlands dominated by oak and pine trees, and vast stretches of heather, which burst into vibrant purple blooms during the summer months. Wildlife enthusiasts will delight in the opportunity to see poisonous adder snakes (a venomous snake species native to the region), wild boar, deer, badgers, foxes, Eurasian eagle-owls, and black woodpeckers. Anthophiles may be able to identify carnivorous sundew plants, orchids, peat and bog moss, and a variety of grasses and lichens. Traversing the network of well-marked hiking and cycling trails, visitors can explore the park's natural wonders at their own pace, be that on horseback, on bike, or on their own two feet. Multiple accommodation options are also available in the park, ranging from tents to chalets with heated pools, for those keen on spending the night in the park.

EUROPE UKRAINE

137 PODILSKI TOVTRY NATIONAL PARK

TO VISIT BEFORE YOU DIE BECAUSE

It is the largest nature conservation area in all of Ukraine.

What sets this park apart is its unique geological formations known as "tovtry" – ancient remnants of coral reefs sculpted over millions of years into towering cliffs, labyrinthine caves, and stunning natural arches. These formations not only provide a spectacular backdrop but also serve as a haven for flora and fauna – more than 1,500 species of plants and 300 species of animals are found within the protected land. That includes the European bison, the symbol of the park. There are also oodles of cultural wonders within the park, including archaeological sites, like ancient forts and settlements, that offer context to the history of the area. Arguably, the best way to experience the park is by hiking or cycling. There are plenty of well-marked trails catering to various difficulty levels.

www.ukraine.com/attractions/national-parks/podilski-tovtry-national-park

EUROPE · UNITED KINGDOM

138 DARTMOOR NATIONAL PARK

TO VISIT BEFORE YOU DIE BECAUSE

A protected area since 1951, this park in Devon contains the largest collection of Bronze Age remains in the United Kingdom.

Notable for its granite rock formations left over from the Carboniferous geologic period, this moorland is rich with wildlife and archaeological artifacts. Dartmoor receives more rain than the surrounding lowlands, so the majority of the land is covered in peat bogs. The area is also subject to high winds and has very acidic soil, and as a result was never intensely farmed, allowing the natural ecosystems to flourish. Visitors can enjoy swimming, picnicking, whitewater rafting, kayaking, rock climbing, and mountain biking. Designated a Special Area of Conservation by the EU's Habitats Directive, Dartmoor hosts hundreds of species of moss, liverwort, and lichen which grow on the exposed granite. The park is known for its large array of bird species, including several species (such as the skylark and common snipe) that have declined in other parts of the UK. Other wildlife common to the area include ponies, shaggy highland cattle, sheep, bats, lizards, rare flies, otters, and Atlantic salmon.

visitdartmoor.co.uk

EUROPE WALES (UK)

139 SNOWDONIA NATIONAL PARK

TO VISIT BEFORE YOU DIE BECAUSE

There's mountains of activities here, from climbing Snowdon to exploring hidden pools, stone monuments, and coastal walks.

Covering 823 square miles, Snowdonia (otherwise known as Eryri) is Wales' largest national park. The area encompasses some of the most dramatic landscapes in the United Kingdom, including Snowdon, the highest peak in Wales, standing tall at 3,559 feet above sea level. From its summit (also accessible via a heritage mountain railway), trekkers can overlook the rolling hills, glittering lakes, and verdant valleys below. Beyond the towering mountain, the park is a patchwork of other ecosystems, from ancient woodlands and meandering rivers to expansive moorland and serene lakes. Amidst it all are centuries-old villages, including some with castles, fortresses, and ruins from a bygone era. For those with an adventurous spirit, Snowdonia offers a vast network of hiking trails, ranging from leisurely strolls to challenging tramps. Rock climbers can scale the sheer cliffs and crags, while mountain bikers can carve their way through winding forest trails.

snowdonia.gov.wales

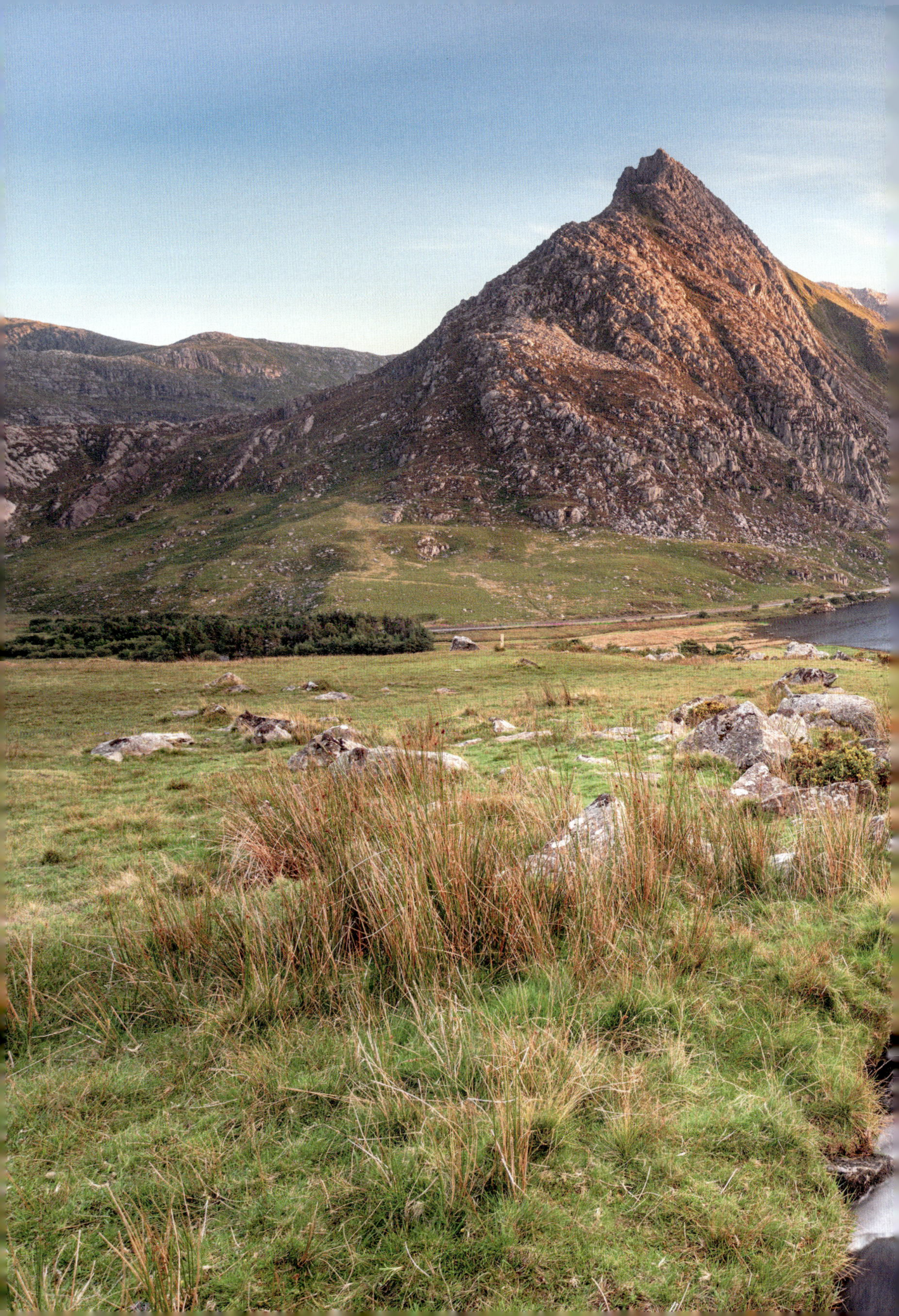

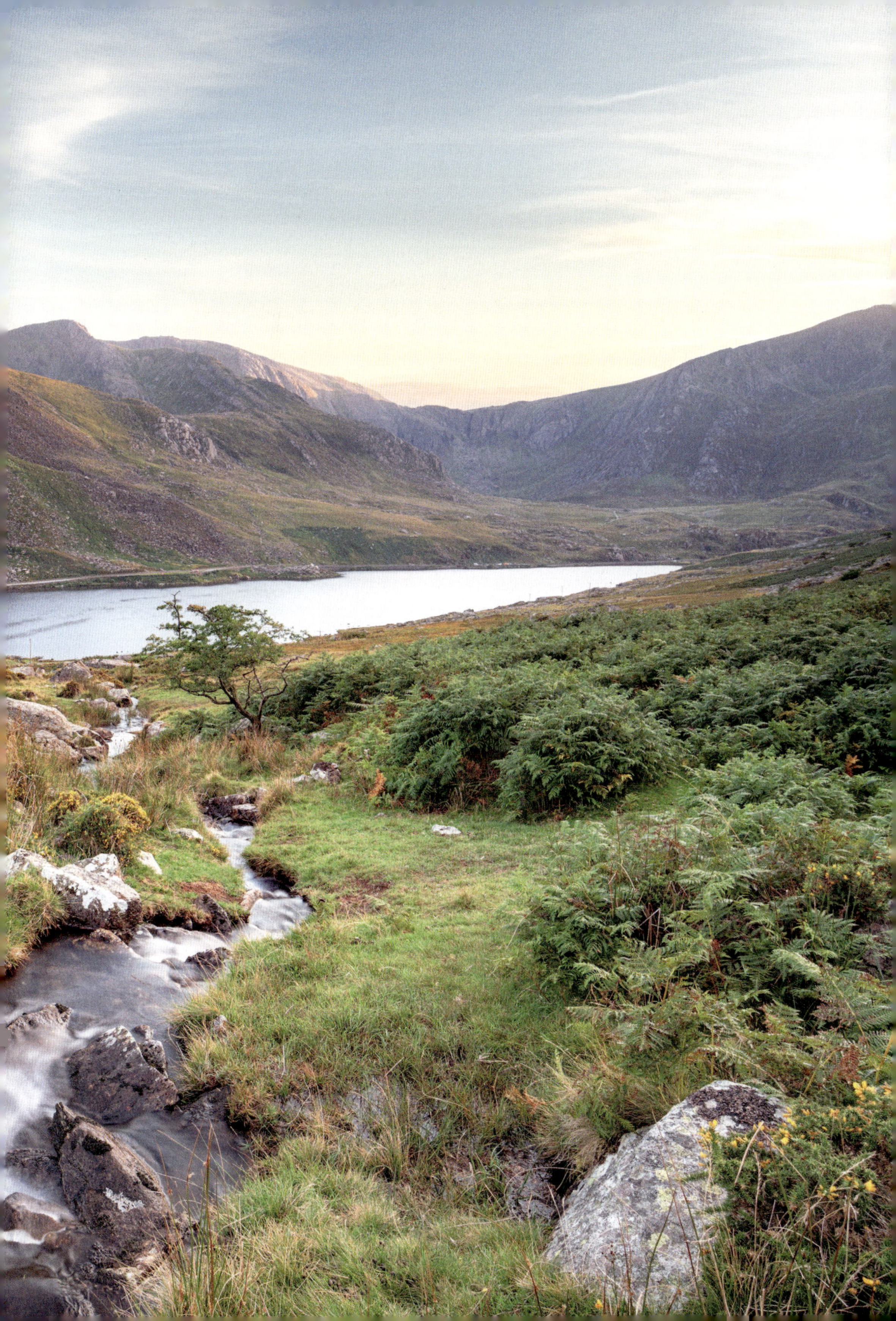

OCEANIA   AUSTRALIA

140 ULURU-KATA TJUTA NATIONAL PARK

TO VISIT BEFORE YOU DIE BECAUSE

You see one of the most important cultural sites of Australia's Aboriginal people, the majestic Uluru, which glows red at dawn and sunset.

Two of the most important geological formations in Australia's rugged outback are found within this park and Unesco World Heritage Site: Uluru (formerly known as Ayers Rock) and Kata Tjuta (also known as Mount Olga). Uluru, a massive sandstone monolith rising 1,142 feet above the surrounding plains, is one of Australia's most recognizable landmarks and one that holds immense spiritual significance to the Indigenous people of the region. Kata Tjuta, which translates to "many heads," is a group of 36 red-rock domes about 15 miles away, which were also key to the traditional belief system of the Anangu Aboriginal people, the original inhabitants of the area (some records suggest they have been there for more than 10,000 years). In addition to the rocky landmarks, the national park is home to myriad desert-adapted animals, including kangaroos, wallabies, emus, and beyond. There are also large numbers of different types of desert wildflowers, grasses, and shrubs scattered throughout the area.

parksaustralia.gov.au/uluru

OCEANIA · AUSTRALIA

141 GREAT BARRIER REEF MARINE PARK

TO VISIT BEFORE YOU DIE BECAUSE

Encompassing the largest reef in the world, this marine park boasts spectacular scenic views above and below the water.

Sir David Attenborough once described the Great Barrier Reef as one of the most beautiful places in the entire world. He's not wrong – the reef is renowned for its stunning array of coral formations, which come in a kaleidoscopic array of colors and shapes. These corals provide a habitat for an extraordinary variety of marine life, including over 1,500 species of fish, ranging from tiny clown fish to massive whale sharks. With more than 2,900 dazzling individual coral reefs and 900 islands stretching across 133,000 square miles, it's the world's biggest single structure made by living organisms. It's so big it can be seen from outer space. Visitors to the park can explore its watery depths by snorkeling, scuba diving, or cruising in a glass bottom boat. Alternatively, helicopter and seaplane tours provide a unique aerial perspective of the reef, showcasing the vastness and beauty of it from above.

www2.gbrmpa.gov.au

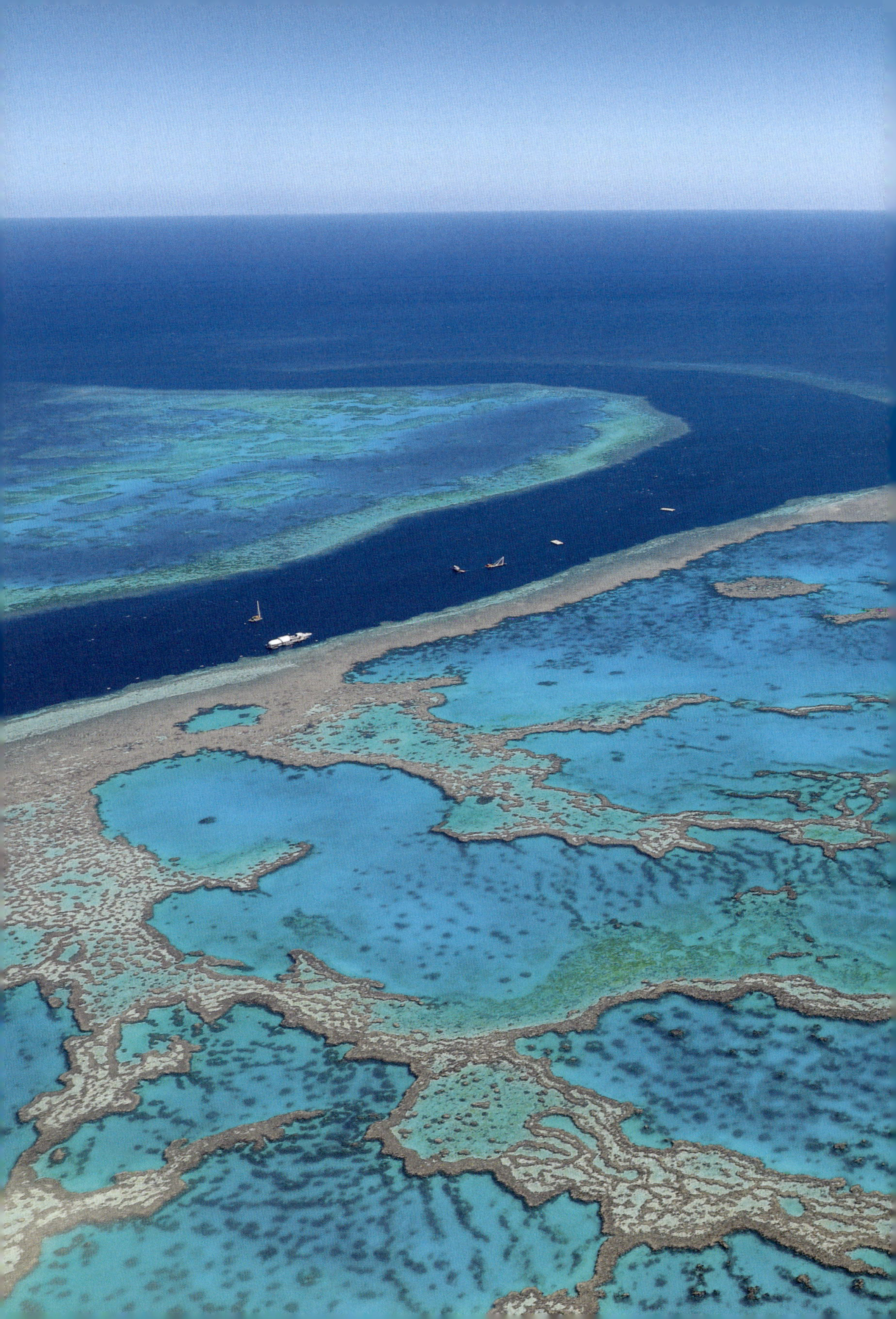

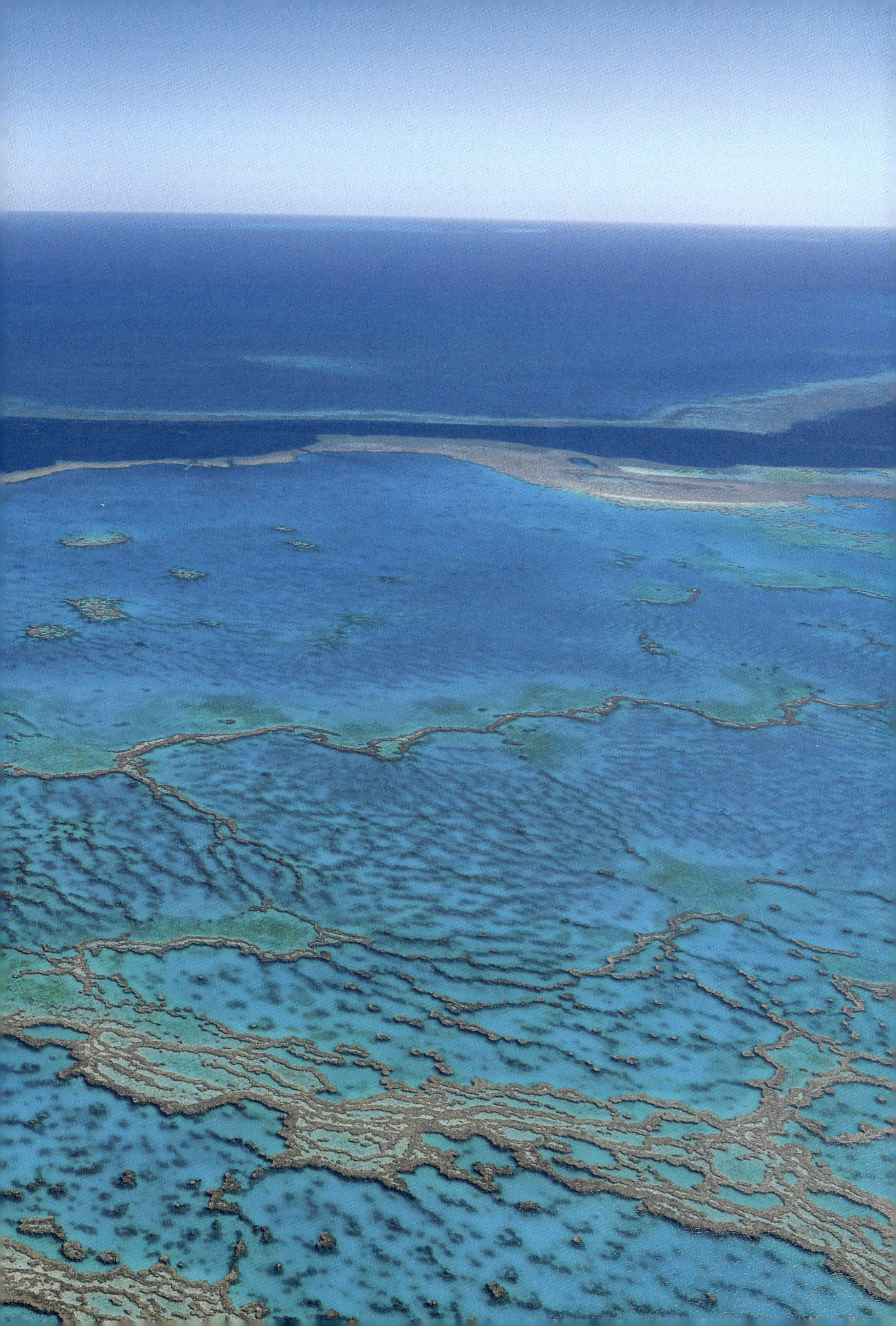

OCEANIA AUSTRALIA

142 CAPE RANGE NATIONAL PARK

TO VISIT BEFORE YOU DIE BECAUSE

Snorkel alongside giant manta rays, explore hundreds of caves, and wonder at one of the world's largest coral reefs at this park in Western Australia.

Did you know that manta rays can grow to have a wingspan of 22 feet? They also can live 50 years and eat roughly 60 pounds of plankton a day. And at Cape Range National Park, you can go snorkeling with these gentle giants. They're far from the only animals worth seeing at this national park located within the Ningaloo Coast World Heritage Area. Other marine animals include sea turtles and humpback whales, which swim amongst one of the largest fringing coral reefs in the world. On land, mammals include emus, dingos, wallabies, kangaroos, and more than 100 species of birds, many of which can be viewed from the variety of hiking trails that weave throughout the ancient gorges and mangrove forests. Interestingly, there are more than 700 caves scattered throughout the park (and likely many more that have yet to be discovered) and upwards of 600 different types of wildflowers.

OCEANIA / AUSTRALIA

143 KAKADU NATIONAL PARK

TO VISIT BEFORE YOU DIE BECAUSE

Though it's the second-largest park in Australia, it's bigger than some countries, and filled with an immense variety of natural landscapes.

Located in Australia's Northern Territory, Kakadu National Park sprawls across more than 7,600 square miles. Within its borders, Kakadu showcases an astonishing array of landscapes, ranging from vast wetlands and lush rainforests to tidal mudflats and monsoon forests. This allows for more than 1,700 plant species and upwards of 280 bird species, 60 mammal species, 50 freshwater species, 110 reptiles, and 10,000 insect species. In addition to its natural beauty, Kakadu is also renowned for its rich cultural heritage. The park is home to ancient Aboriginal rock art sites, some of which date back thousands of years and depict hunting and gathering practices, social structures, and ritual ceremonies. These intricate artworks provide a window into the spiritual and cultural beliefs of the traditional owners of the land and help prove that the area has been continuously inhabited for at least 40,000 years.

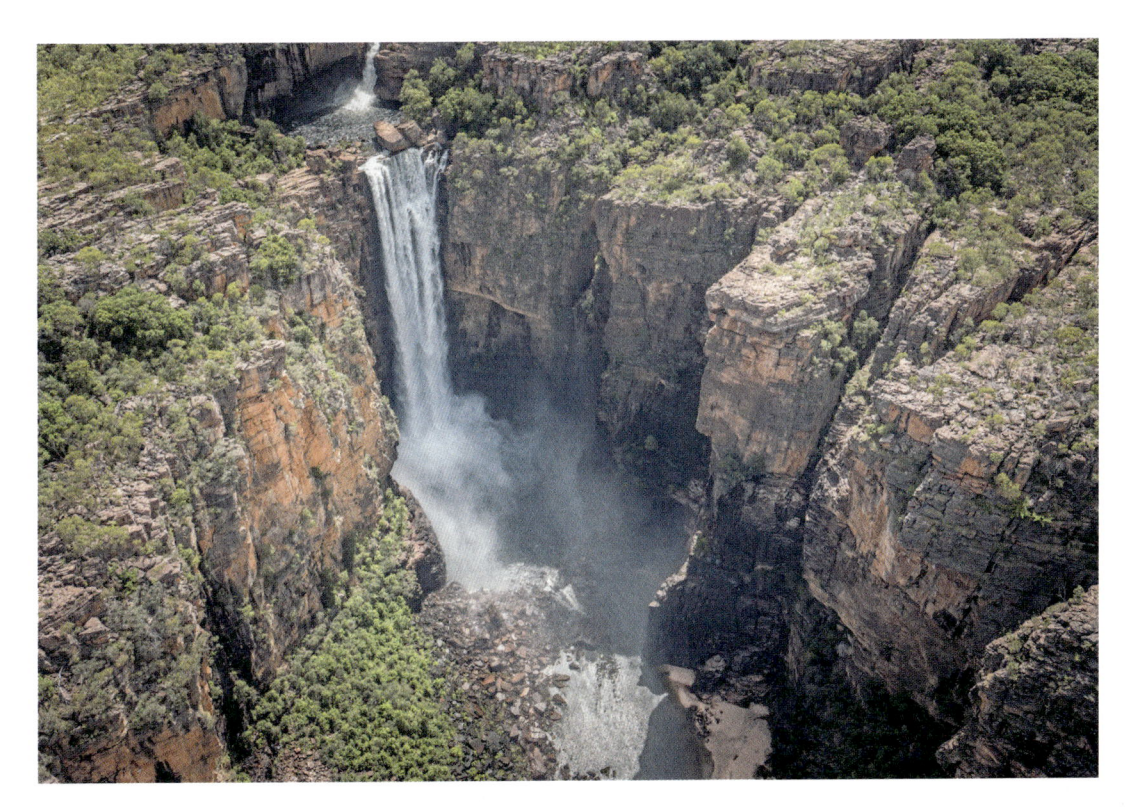

parksaustralia.gov.au/kakadu

OCEANIA   AUSTRALIA

144 LEEUWIN-NATURALISTE NATIONAL PARK

TO VISIT BEFORE YOU DIE BECAUSE

This popular coastal park located just south of Perth is known for its surf breaks, dunes, lush forests, and limestone cliffs.

Located on the picturesque southwestern coast of Western Australia, Leeuwin-Naturaliste National Park is named for the lighthouses on the capes at either end of the park. Meandering trails wind through the park, inviting exploration and immersion in nature. There are karri and jarrah forests (two types of exceptionally tall trees in the eucalyptus family), whose canopies provide shelter to an assortment of native birdlife, such as the red-winged fairy-wren and white-naped honeyeater, and mammals, like possums, wallabies, kangaroos, and the southern brown bandicoot. Venturing towards the coast unveils sheer limestone cliffs that overlook the vast expanse of the Indian Ocean. The roughly 75 miles of coastline at Leeuwin-Naturaliste National Park is also dotted with pristine beaches of golden sands, an idyllic setting for beachcombers, surfers, and sun-seekers alike, while the waters teem with biodiversity, including humpback whales, dolphins, sting rays, and colorful reef fish.

exploreparks.dbca.wa.gov.au/park/leeuwin-naturaliste-national-park

OCEANIA FIJI

145 BOUMA NATIONAL PARK

TO VISIT BEFORE YOU DIE BECAUSE

This true island paradise comes with unbelievably verdant rainforests filled with rare birds and waterfalls.

Bouma National Park is a breathtaking natural reserve encompassing roughly 80 percent of the island of Taveuni in Fiji. The park boasts some incredible flora (Taveuni is nicknamed "the Garden Island" thanks to its abundant and lush plant life), including the crimson and white tagimoucia flower, which is the symbol of the nation. There are also more than 100 species of birds, such as silktails, orange doves, and kula lorikeets, and perhaps your best chance of seeing them is on the Lavena Coastal Walk, which traces the coast for three miles each way. Trickling streams and cascading waterfalls punctuate the landscape, adding to the park's serene ambiance. One of the park's most iconic attractions is the Tavoro Waterfalls, a series of three stunning cascades, each with a natural pool, set deep in the rainforest. There are four villages within the park, where travelers can sign up to go for guided walks or snorkeling off traditional rafts.

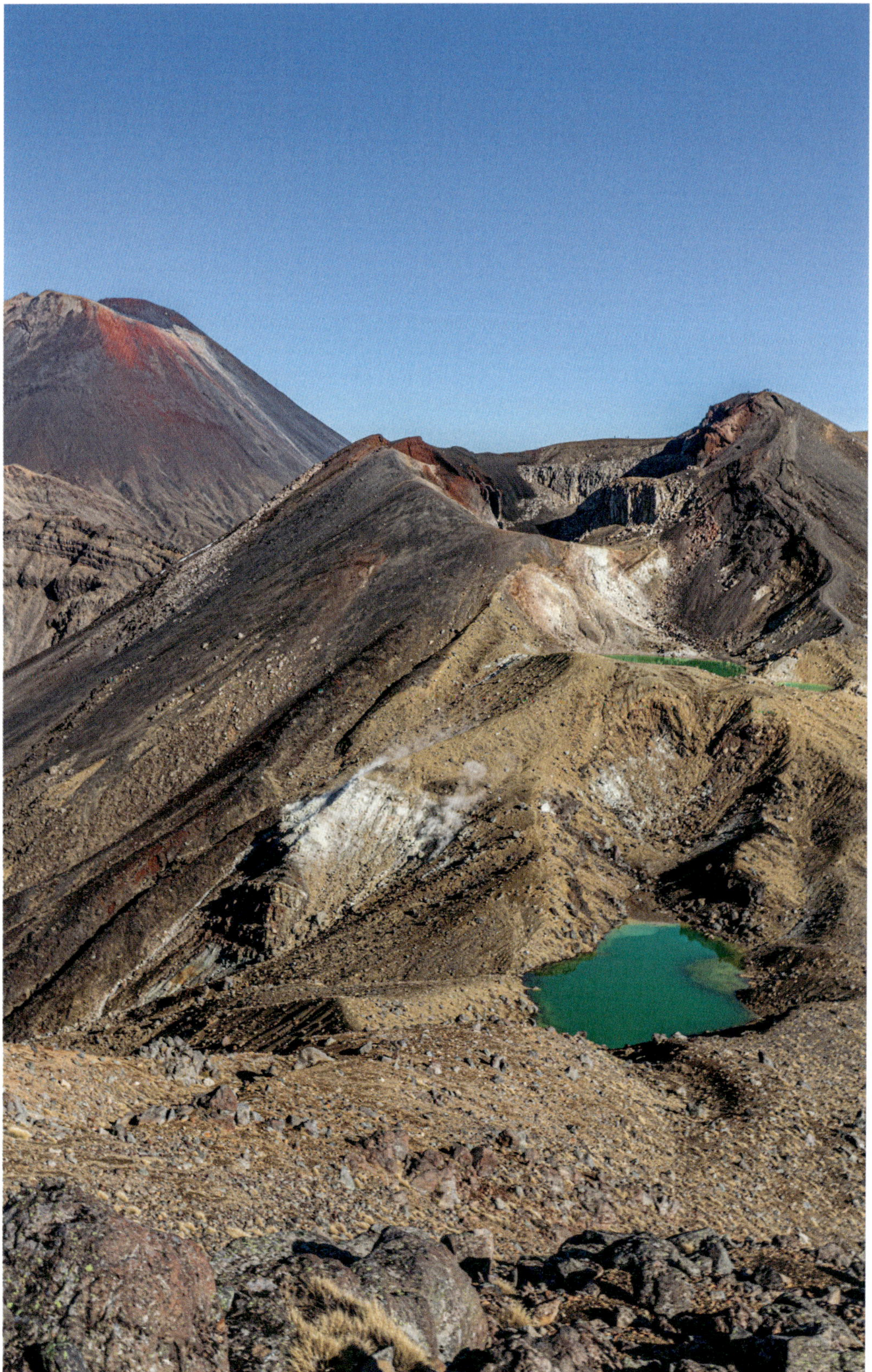

OCEANIA NEW ZEALAND

146 TONGARIRO NATIONAL PARK

TO VISIT BEFORE YOU DIE BECAUSE

Volcanic landscapes and Māori culture are on full display at this 300-square-mile park which has attracted adventurers since 1887.

Located at the heart of New Zealand's North Island, this park encompasses three active volcanoes: Tongariro, Ngauruhoe, and Ruapehu. The trio of peaks, with their roughly cut slopes and steaming vents, create a dramatic backdrop for outdoor adventures and exploration. When it was established in 1887, it was only the sixth national park to be named. Now, one of the most popular ways to experience the land is by trekking. One of the park's most famous hiking trails is the Tongariro Alpine Crossing, a challenging nearly 12-mile trek that traverses volcanic craters, glacial valleys, and lunar-like landscapes. It is widely considered one of the best single-day hikes on the planet and one of New Zealand's nine Great Walks. Tongariro National Park is also steeped in Māori culture and tradition, with sacred sites and cultural landmarks scattered throughout the landscape. The park's Māori name, Te Pokai-o-Wherua, translates to "The Bosom of Earth Mother," reflecting spiritual significance to the Indigenous people.

nationalpark.co.nz

OCEANIA NEW ZEALAND

147 FIORDLAND NATIONAL PARK

TO VISIT BEFORE YOU DIE BECAUSE

Enjoy three of the best hikes in New Zealand and the chance to see rare birds.

As its name might suggest, Fiordland National Park is characterized by its deep, glacier-carved fjords, including the iconic Milford Sound, Doubtful Sound, and Dusky Sound. It also includes some of New Zealand's highest peaks, such as Mitre Peak and Mount Tutoko, which are often cloaked in snow, creating a striking contrast against the verdant forests and deep blue waters below. The park is also home to a variety of wildlife, including native birds like the kea (New Zealand's mischievous alpine parrot), the takahe (a flightless bird once thought to be extinct), and the tawaki (otherwise known as the Fiordland crested penguin, which is the rarest type of penguin in the world). The Milford Track, one of the most famous hiking trails in the world, takes adventurers on a 33-mile journey through some of the park's most spectacular landscapes, from lush temperate forests to alpine passes. It's one of New Zealand's 10 Great Walks – two others, the Routeburn Track and the Kepler Track, are also within the park boundaries.

fiordland.org.nz/visit/fiordland-national-park

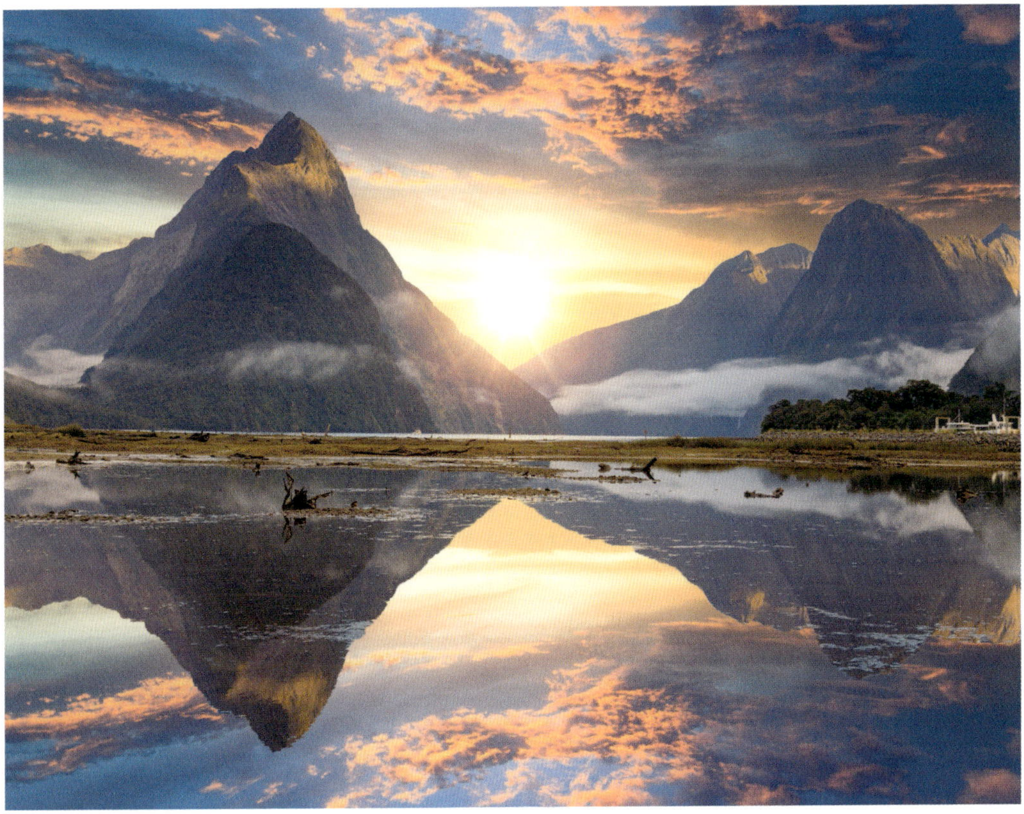

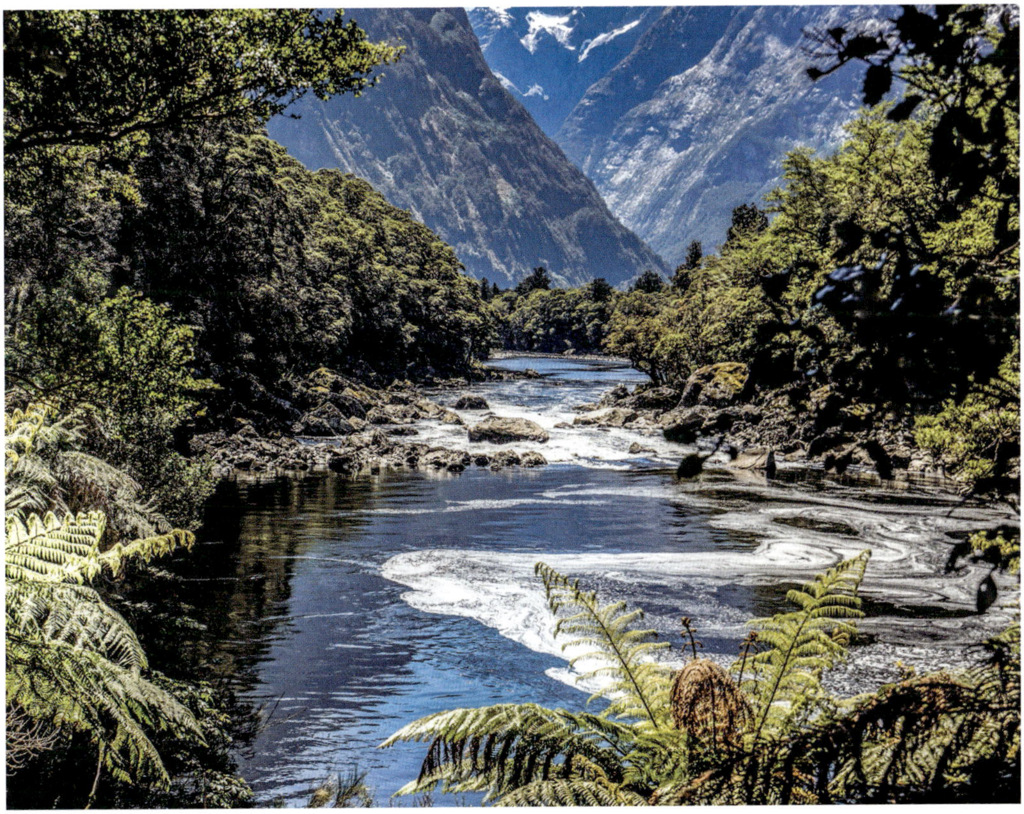

OCEANIA NEW ZEALAND

148 AORAKI/MOUNT COOK NATIONAL PARK

TO VISIT BEFORE YOU DIE BECAUSE

A rugged land of ice and rock, this park comprises 19 peaks over 9,800 feet, including New Zealand's highest mountain, Aoraki.

Eponymously named for the tallest mountain in New Zealand, Aoraki (meaning "Cloud Piercer" in Māori) or Mount Cook reaches an impressive height of 12,218 feet. Like other great peaks, its snow-capped summit is often shrouded in mist and clouds, adding to its mystique and grandeur (the Māori consider the Aoraki to be the most sacred of the ancestors they descended from). The mountain isn't the only superlative Aoraki/Mount Cook National Park possess, though. The protected land also boasts the mighty Tasman Glacier, the longest glacial mass in the country. With numerous hiking trails catering to varying skill levels, the park is a haven for outdoor enthusiasts. Whether you're trekking to Hooker Lake to see the mountain reflected in its turquoise waters or embarking on the challenging Mueller Hut Route for panoramic vistas of the Southern Alps, there's awe-inspiring scenery at every turn. And who knows, perhaps you'll see the kea, the only alpine parrot in the world, or the takahe, a large, rare flightless bird, along the way.

newzealand.com/us/feature/national-parks-aoraki-mount-cook

OCEANIA NEW ZEALAND

149 ABEL TASMAN NATIONAL PARK

TO VISIT BEFORE YOU DIE BECAUSE

Immerse yourself in the beauty of New Zealand's coastal wilderness at this park renowned for its golden beaches and sculptural granite cliffs.

Abel Tasman National Park, nestled at the northwestern tip of New Zealand's South Island, is a pristine coastal paradise renowned for its golden beaches, turquoise waters, and lush native forests. Established in 1942 and named after the Dutch explorer Abel Tasman, the first European to sight New Zealand in 1642, it is the country's smallest national park. However, its coastal location provides a unique blend of marine and terrestrial ecosystems, making it rich in diverse fauna, ranging from fur seals to the kakariki, a colorful native parakeet species. It's also an exciting place for outdoor adventures. Hikers might consider tackling the famous Abel Tasman Coast Track, one of New Zealand's Great Walks, which winds its way along the shoreline for 38 miles. Or they could spend time in and on the water by swimming, snorkeling, or sea kayaking in the sheltered bays. The park's warm climate and tranquil waters make it a lovely destination for water-based adventures year-round.

doc.govt.nz/parks-and-recreation/places-to-go/nelson-tasman

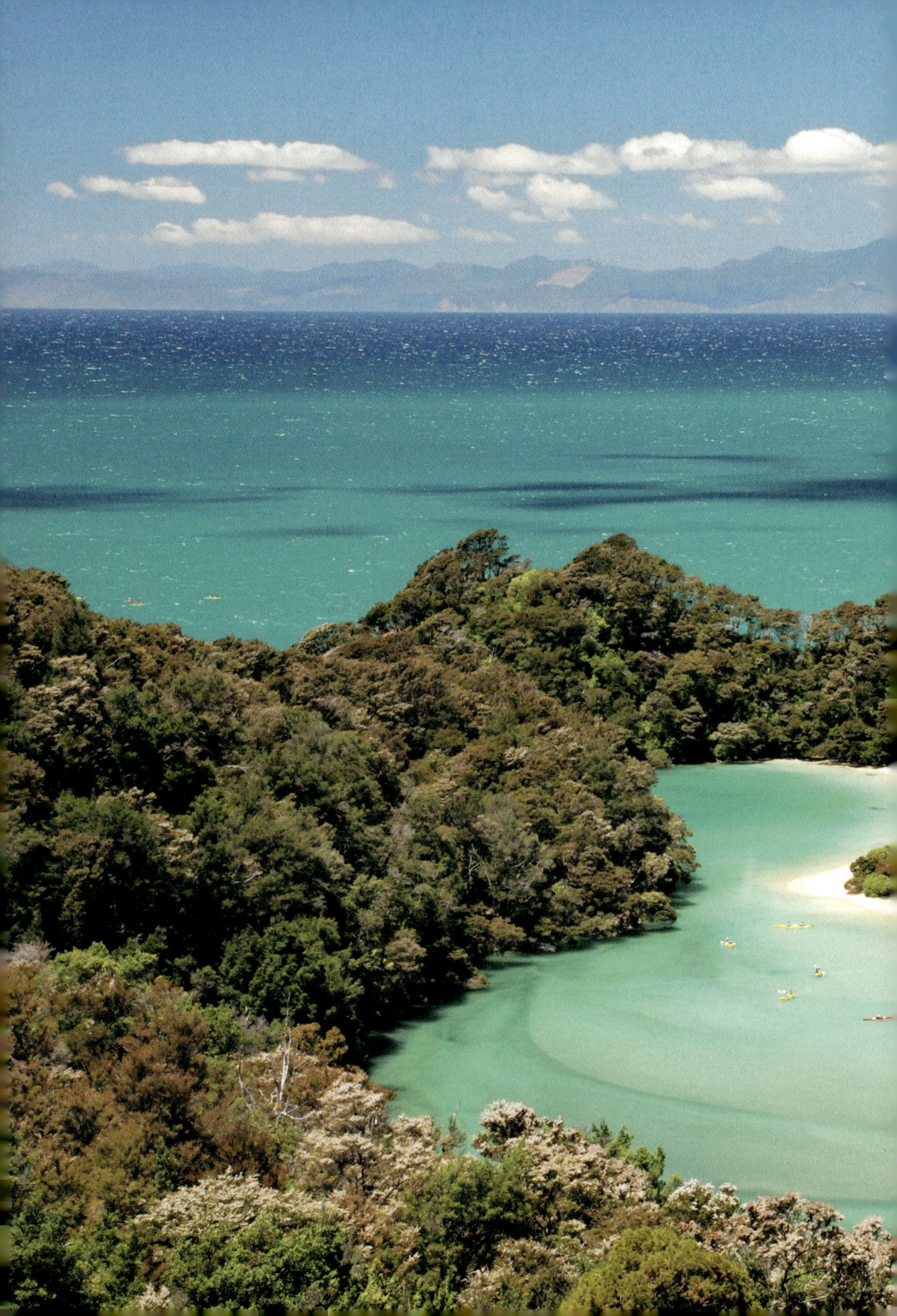

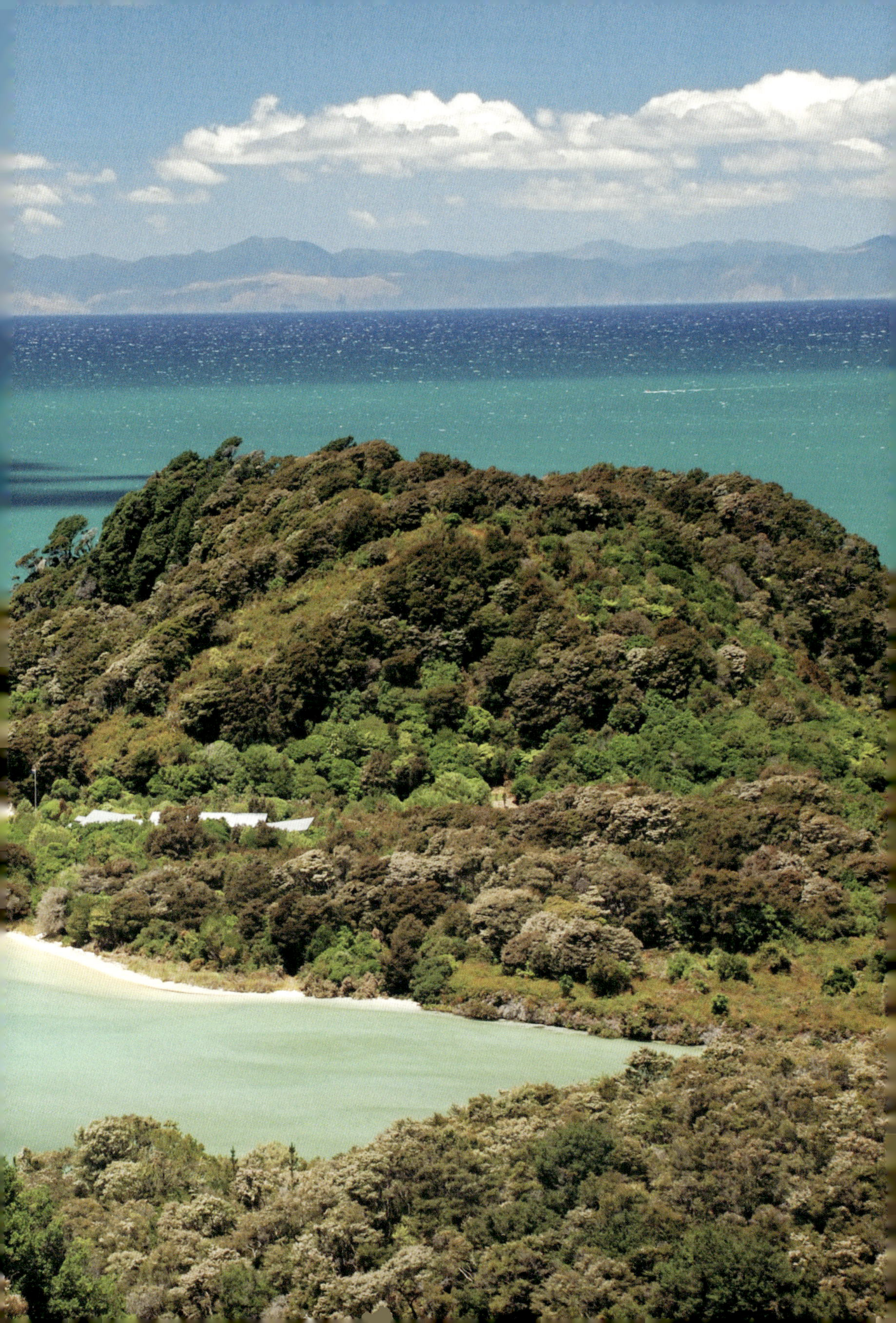

OCEANIA NEW ZEALAND

150 WHANGANUI NATIONAL PARK

TO VISIT BEFORE YOU DIE BECAUSE

Here you can find the first river to be legally recognized as a person.

In 2017, the Whanganui River on New Zealand's North Island became the world's first river to be granted legal personhood, concluding a 140-year discourse on the matter. This groundbreaking legislation acknowledged the profound spiritual bond between the river and the Māori iwi, who consider it an ancestral entity. Stretching 180 miles from Mt Tongariro to the Tasman Sea, the Whanganui River stands as the longest navigable waterway in New Zealand. It forms a prominent route for intrepid travelers engaging in one of New Zealand's esteemed Great Walks, offering opportunities for exploration by canoe, kayak, jetboat, and bike. Along the river, majestic waterfalls cascade from moss-covered cliffs, while hidden gorges reveal secrets of ancient geological history. Along the riverbanks, remnants of Māori settlements offer glimpses into the region's rich cultural heritage, with fortified villages (pā) and traditional dwelling sites (kainga) bearing witness to centuries of human presence.

doc.govt.nz/parks-and-recreation/places-to-go/manawatu-whanganui

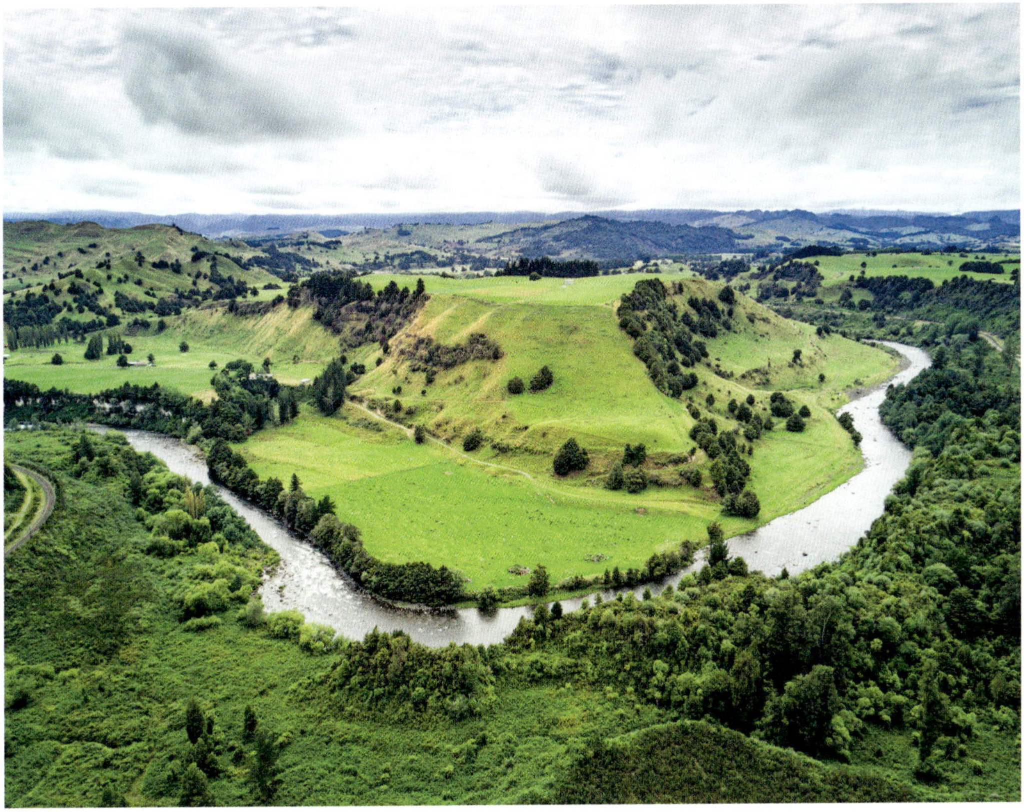

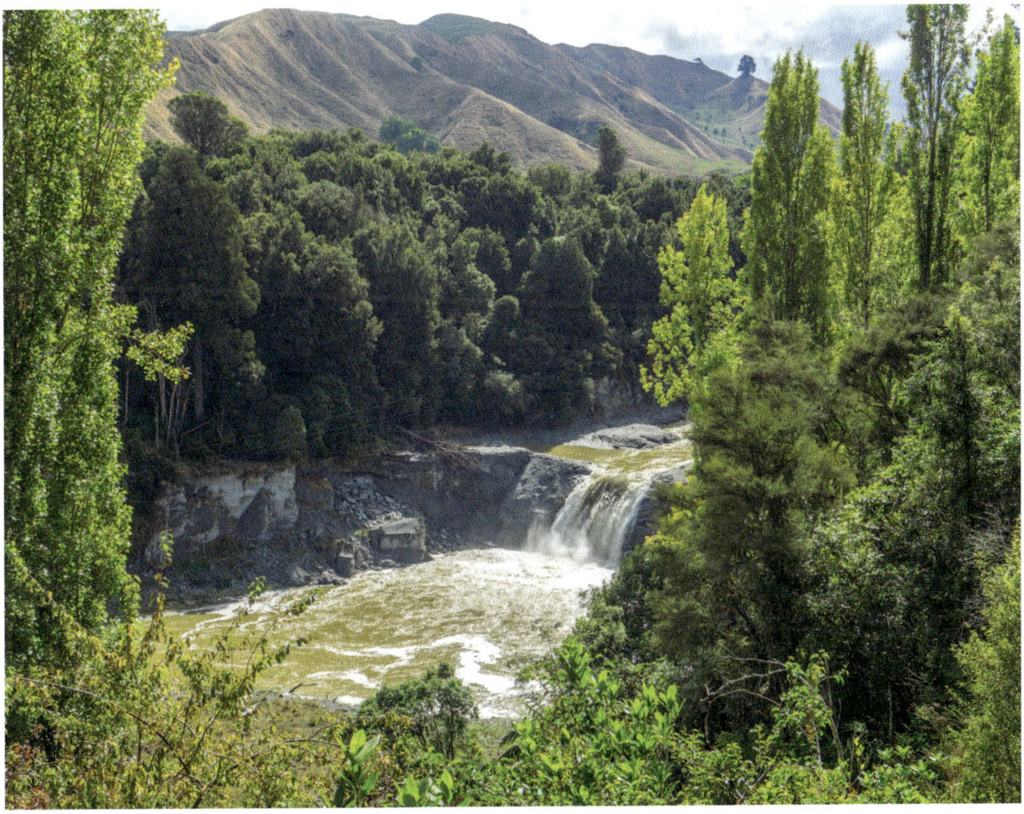

Index 150 National parks

Abel Tasman National Park 247
Abisko National Park 223
Acadia National Park 132
Altyn-Emel National Park. 160
Amboró National Park 58
Aoraki/Mount Cook
  National Park 246
Aparados da Serra National Park . . 64
Arches National Park 129
Arikok National Park 57
Badaling National Forest Park 146
Banff National Park 73
Big Bend National Park 121
Bouma National Park 241
Butrint National Park. 182
Bwindi Impenetrable
  National Park 41
Cairngorms National Park. 219
Canaima National Park 135
Cape Range National Park. 238
Cat Ba National Park 179
Channel Islands National Park. 134
Chapada Diamantina
  National Park 60
Cheile Nerei-Beusnita
  National Park 218
Chitwan National Park 163
Chobe National Park 11
Cinque Terre National Park. 207
Connemara National Park. 203
Corcovado National Park 80
Cotopaxi National Park 84
Dartmoor National Park 230
De Meinweg National Park 228
Denali National Park 120
Dolomiti Bellunesi National Park . 204
Etosha National Park 22
Fernando de Noronha Marine
  National Park 67
Fiordland National Park 244
Forillon National Park 70
Fuji-Hakone-Izu National Park 154
Galapágos National Park 81
Gauja National Park 210
Glacier National Park 109
Göreme National Park 176
Gorkhi-Trelj National Park 161
Gouraya National Park 10
Gran Paradiso National Park 206
Grand Canyon National Park 110
Grand Teton National Park 115
Great Barrier Reef Marine Park . . . 235
Great Smoky Mountains
  National Park 116
Guilin and Lijiang River
  National Park 147
Gunung Mulu National Park. 143
Haleakalā National Park 117
Hawai'i Volcanoes National Park. . .118
Hell's Gate National Park 17
Het Nationale Park
  De Hoge Veluwe 225
Hoge Kempen National Park 185
Hossa National Park. 191
Hundred Islands National Park . . . 164

Hwange National Park 46
Iguazú National Park 51
Isalo National Park 19
Jasper National Park 69
Jigme Dorji National Park. 138
Jirisan National Park 168
Joshua Tree National Park 108
Jostedalsbreen National Park. 211
Jotunheimen National Park 214
Kaieteur National Park 86
Kakadu National Park 239
Katmai National Park. 112
Kenai Fjords National Park. 113
Khao Sok National Park. 173
Kilimanjaro National Park 39
Kinabalu National Park 139
Komodo National Park. 158
Krka National Park 187
Kruger National Park 31
Lake Nakuru National Park 18
Lambir Hills National Park 142
Leeuwin-Naturaliste
  National Park 240
Lençóis Maranhenses
  National Park 66
Los Glaciares National Park 53
Manu National Park 90
Masada National Park 150
Mount Rainier National Park 92
Naejangsan National Park. 167
Namib-Naukluft National Park 23
Nationaal Park de Biesbosch. 224
National Park Langue de Barbarie. . .30
New River Gorge National Park . . . 94
Ngorongoro National Park 40
Niah National Park. 141
Northeast Greenland
  National Park 198
Odzala-Kokoua National Park 24
Olympus National Park 195
Ordesa and Monte Perdido
  National Park 222
Oulanka National Park. 192
Peneda-Gerês National Park. 216
Petrified Forest National Park 106
Phong Nha-Kẻ Bàng
  National Park 178
Plitvice Lakes National Park. 190
Podilski Tovtry National Park 229
Preah Monivong Bokor
  National Park 144
Ranthambore National Park. 153
Rapa Nui National Park 77
Redwood National Park. 122
Réunion National Park 25
Rila National Park. 186
Rocky Mountain National Park 125
Sagarmartha National Park. 162
Saxon Switzerland National Park . 194
Seoraksan National Park. 166
Sequoia and Kings Canyon
  National Park 126
Serengeti National Park 35
Shenandoah National Park 128
Simien Mountains National Park . . .16

Skeleton Coast National Park 20
Slovak Karst National Park 220
Snowdonia National Park 231
South Luangwa National Park 44
Stolby Nature Reserve 165
Sundarbans National Park. 152
Table Mountain National Park 34
Taroko National Park. 172
Tatra National Park. 215
Tierra del Fuego National Park 52
Tikal National Park 85
Tongariro National Park 243
Torres del Paine National Park 74
Tortuguero National Park 78
Triglav National Park. 221
Tulum National Park 87
Uluru-Kata Tjuta National Park. . . 234
Vatnajökull National Park. 199
Victoria Falls National Park 47
Virunga National Park 12
Volcan Baru National Park 89
Volcanoes National Park 28
Voyageurs National Park 95
Wadi El Gemal National Park 14
Wadi Rum Protected Area. 155
Wadi Wurayah National Park 177
Whanganui National Park. 250
White Desert National Park 15
White Sands National Park 98
Wrangell-St.Elias National Park . . . 99
Yala National Park 169
Yellowstone National Park 102
Yosemite National Park 104
Zhangye Danxia National Geopark .145
Zion National Park 105

© Photos

Park 1: Wirestock / Park 2: estivillml / Park 3: (left): Photocech, (right): guenterguni / Park 4: Mohamed Ramez / Park 5: Wirestock / Park 6: helovi / Park 7: WLDavies / Park 8: Anna_Om / Park 9: dennisvdw / Park 10: (left): namibelephant, (right): zulufriend / Park 11: paulafrench / Park 12: 2630ben / Park 13: guenterguni / Park 14: (1): Infografick, (2): Alexandra Bellekens / Park 15: (left): guenterguni, (right, top): stellalevi, (right, bottom): Dennis Stogsdill / Park 16: Безгодов / Park 17: (1): Moonstone Images (2): SL_Photography / Park 18: kavram / Park 19: (1): MrRuj (2): pchoui / Park 20: (left, top): 1001slide, (left, bottom): Byrdyak, (right): hadynyah / Park 21: SimonSkafar / Park 22: (1): Pedro Ferreira do Amaral, (2): ANDREYGUDKOV / Park 23: (left): JeremyRichards, (right, top): JeremyRichards, (right, bottom): Bobbushphoto / Park 24: paulafrench / Park 25: (1): 2630ben, (2): 3dan3 / Park 26: (left): David Aguilar Photography, (right): kavram / Park 27: Fyletto / Park 28: (1): Serjio74, (2): ailtonsza / Park 29: (left): Jan-Schneckenhaus, (right): boblin / Park 30: (left): Matthias Kestel, (right, top): hernan4429, (right, bottom): Elijah-Lovkoff / Park 31: (left & right): diegograndi, (2): diegograndi / Park 32: (left): diegograndi, (right, top & bottom): diegograndi / Park 33: MaRabelo / Park 34: cacio murilo de vasconcelos / Park 35: (left, top): lucky-photographer, (left, bottom): GlowingEarth, (right): MJ_Prototype / Park 36: (left) Damien VERRIER, (right): Luc Rousseau / Park 37: (left): YinYang, (right): wwing / Park 38: (left): MBPROJEKT_Maciej_Bledowski, (right): franckreporter / Park 39: (left): f11photo, (right): tankbmb / Park 40: (left): Francesco Ricca Iacomino, (right, top): Eefje Verbeek, (right, bottom): jarnogz / Park 41: Eisenlohr / Park 42: (1): guenterguni, (2): DC_Colombia / Park 43: PatricioHidalgoP / Park 44: diegograndi / Park 45: Wirestock / Park 46: LanaCanada / Park 47: (left): NTCo, (right): Rainer Lesniewski / Park 48: (left): Marek Stefunko, (right): OSTILL / Park 49: (left): Drew Payne, (right): Wirestock / Park 50: Different_Brian / Park 51: (1): StevenSchremp, (2): BlueBarronPhoto / Park 52: Jim Brown / Park 53: (1): Cavan Images, (2): Wirestock / Park 54: (left): tiny-al, (right): Megan Brady / Park 55: Paul D Wade / Park 56: Poul Riishede / Park 57: (left): sanfel, (right, top): Jeffrey Ross, (right, bottom): Bonnie Nordling / Park 58: garytog / Park 59: HaizhanZheng / Park 60: (left): jose1983, (right): kojihirano / Park 61: sarkophoto / Park 62: Jaime Espinosa de los Monteros / Park 63: (left): christiannafzger, (right): Brian Evans / Park 64: KenCanning / Park 65: Don White / Park 66: (left): theartist312, (right, top): Ken McCurdy, (right, bottom): joebelanger / Park 67: davidhoffmannphotography / Park 68: dhughes9 / Park 69: (left): franckreporter, (right): lucky-photographer / Park 70: (left): Douglas Rissing, (right): Matt Dirksen / Park 71: (left): Andrelix, (right): Juan R Garcia / Park 72: AppalachianViews / Park 73: (1): f11photo, (2): Nancy C. Ross / Park 74: (left): Harry Collins, (right, top): AppalachianViews, (right, bottom): James Griffiths Photography / Park 75: kellyvandellen / Park 76: (1): mariusz_prusaczyk, (2): nicolasdecorte / Park 77: slowmotiongli / Park 78: yusnizam / Park 79: (left, top): Juhku, (left, bottom): lillitve, (right): Darren Creighton / Park 80: davincidig / Park 81: zodebala / Park 82: undefined undefined / Park 83: wonry / Park 84: zhengzaishuru / Park 85: (1): luxiangjian4711, (2): aphotostory / Park 86: (left): Mindaugas Dulinskas, (right): jsteck / Park 87: Ekaterina Aleshinskaya / park 88: zodebala / Park 89: magicflute002 / Park 90: (1): SzymonBartosz, (2): SzymonBartosz / Park 91: (left): ksumano, (right): donnchans / Park 92: Ozbalci / Park 93: ShevchenkoAndrey / Park 94: fotoVoyager / Park 95: Si Mon / Park 96: Alexpunker / Park 97: Pavel Sipachev / Park 98: tawatchaiprakobkit / Park 99: Atakorn / Park 100: sueuy song / Park 101: (1): Sachin Fernando, (2): T_o_m_o / Park 102: PonAek / Park 103: (1): bloodua, (2): southtownboy / Park 104: Elenastudio / Park 105: Jeff Kingma / Park 106: vinhdav / Park 107: (1): fbxx, (2): melis82 / Park 108: (left): DC_Colombia, (right): LeszekCzerwonka / park 109: (left, top): Eefje Verbeek, (left, bottom): kim Willems, (right): Eefje Verbeek / Park 110: Nadya85 / Park 111: (1 &2): DaLiu / Park 112: supergenijalac / Park 113: VSFP / Park 114: (left): artenex, (right, top): Karl Ander Adami, (right, bottom): artenex / Park 115: Asergieiev / Park 116: (1): frantic00, (2): StuartDuncanSmith / Park 117: XXXX / Park 118: (1): takepicsforfun, (2): Sergdid / Park 119: (left, top): Louis-Michel DESERT, (left, bottom): AlbertMi, (right): AlbertMi / Park 120: (left): Emanuela Seppi, (right): Orietta Gaspari / Park 121: ueuaphoto / Park 122: (1): Sean Pavone, (2): bennymarty / Park 123: imantsu / Park 124: (1): HildaWeges, (2): destillat / Park 125: Eswaran Arulkumar / Park 126: Gosiek-B / Park 127: (left): Sergey_Peterman, (right, top): Vertigo3d, (right, bottom): Sebastian Sonnen / Park 128: heckepics / Park 129: lucentius / Park 130: sedmak / Park 131: gevision / Park 132: elxeneize / Park 133: rusm / 134: XXXX / Park 135: (1): arturasker, (2): Milos Ruzicka / Park 136: JStuij / Park 137: Sergnester / Park 138: ianwool / Park 139: (1): Will Dale, (2): Helen Hotson / Park 140: FiledIMAGE / Park 141: (1): Nautilus Creative, (2): Luisa Trescher Photos / Park 142: Paula Jones / Park 143: JanelleLugge / Park 144: NeilJB / Park 145: Donyanedomam / Park 146: (left): Mark Fitzsimons, (right): DoraDalton / Park 147: (left): Sasithorn Phuapankasemsuk, (right, top): primeimages, (right, bottom): Mark Fitzsimons / Park 148: Kuntalee Rangnoi / Park 149: (1): ChristianB, (2): jfoltyn / Park 150: (left): Uwe Moser, (right, top): LaSalle-Photo, (right, bottom): rumboalla

- 150 | SPAS | YOU NEED TO VISIT BEFORE YOU DIE
- 150 | VINEYARDS | YOU NEED TO VISIT BEFORE YOU DIE
- 150 | BOOKSTORES | YOU NEED TO VISIT BEFORE YOU DIE
- 150 | GARDENS | YOU NEED TO VISIT BEFORE YOU DIE
- 150 | BARS | YOU NEED TO VISIT BEFORE YOU DIE
- 150 | RESTAURANTS | YOU NEED TO VISIT BEFORE YOU DIE
- 150 | WINE BARS | YOU NEED TO VISIT BEFORE YOU DIE
- 150 | HOUSES | YOU NEED TO VISIT BEFORE YOU DIE
- 150 | GOLF COURSES | YOU NEED TO VISIT BEFORE YOU DIE
- 150 | HOTELS | YOU NEED TO VISIT BEFORE YOU DIE

IN THE SAME — SERIES —

150 spas you need to visit before you die
isbn 978 94 014 9747 3
—

150 vineyards you need to visit before you die
isbn 978 94 014 8546 3
—

150 bookstores you need to visit before you die
isbn 978 94 014 8935 5
—

150 gardens you need to visit before you die
isbn 978 94 014 7929 5
—

150 bars you need to visit before you die
isbn 978 94 014 4912 0
—

150 wine bars you need to visit before you die
isbn 978 94 014 8622 4
—

150 houses you need to visit before you die
isbn 978 94 014 6204 4
—

150 golf courses you need to visit before you die
isbn 978 94 014 8195 3
—

150 hotels you need to visit before you die
isbn 978 94 014 5806 1
—

150 restaurants you need to visit before you die
isbn 978 94 014 9570 7

Colophon

Text
Bailey Berg

Book Design
ASB (Atelier Sven Beirnaert)

Typesetting
Keppie & Keppie

Back Cover Image
© Franckreporter

Sign up for our newsletter with news about
new and forthcoming publications on art,
interior design, food & travel, photography and
fashion as well as exclusive offers and events.
If you have any questions or comments about
the material in this book, please do not hesitate
to contact our editorial team: art@lannoo.com

© Lannoo Publishers, Belgium, 2024
D/2024/45/435 - NUR 450/500
ISBN 978-94-014-1970-3

www.lannoo.com

All rights reserved. No part of this publication may
be reproduced or transmitted in any form or by any
means, electronic or mechanical, including photocopy,
recording or any other information storage and
retrieval system, without prior permission in writing
from the publisher.

Every effort has been made to trace copyright holders.
If, however, you feel that you have inadvertently been
overlooked, please contact the publishers.